T0285016

THE
OYSTER
BOOK

*A Chronicle of the World's Most Fascinating
Shellfish—Past, Present, and Future*

DAN MARTINO

A SURREY BOOK

AGATE

CHICAGO

Printed in September 2024

10 9 8 7 6 5 4 3 2 1   24 25 26 27 28

Author photo by Dan Martino
Illustrations by Ievgeniia Lytvynovych unless otherwise noted.

Library of Congress Cataloging-in-Publication Data

Names: Martino, Dan, author.
Title: The oyster book : A chronicle of the world's most fascinating shellfish—past, present, and future / Dan Martino.
Description: Chicago : A Surrey Book, Agate, [2024] | Includes
    bibliographical references. |
Identifiers: LCCN 2024017891 (print) | LCCN 2024017892 (ebook) | ISBN
    9781572843424 (hardcover) | ISBN 9781572848887 (ebook)
Subjects: LCSH: Oysters--History. | Oysters--Conservation. | Animals and
    civilization.
Classification: LCC QL430.7.O9 M37 2024 (print) | LCC QL430.7.O9
(ebook)
    | DDC 594/.4--dc23/eng/20240430
LC record available at https://lccn.loc.gov/2024017891
LC ebook record available at https://lccn.loc.gov/2024017892

Surrey Books is an imprint of Agate Publishing. Agate books are available in bulk at discount prices. For more information, visit agatepublishing.com.

Dedicated to my amazing wife, Laura, and my awesome children,
Adriana, Julian, and Sofia Martino.

Pursue your interests and your life will be interesting.
And have fun.

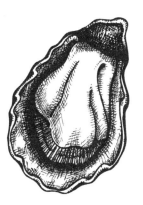

CONTENTS

INTRODUCTION

HUMANS BEGAN TO FARM THE FERTILE CRES-
cent about twelve thousand years ago. The advent of
agriculture created a pillar for modern society to stand
on, allowing for the expansion and proliferation of humans across
the Earth.

While twelve thousand years may seem like a long time, in
actuality it is just a blink compared to the hundreds of thousands
of years of human evolution, during which the human diet consisted mostly of ocean organisms. In fact, we've been eating from
the ocean for so long that humans can consume almost any ocean
species raw and uncooked. From fish to seaweeds to shellfish, the
human digestive system is happy and at home with uncooked ocean
foods; however, even the faintest hint of raw chicken or pork will
surely end in a near-fatal experience.

After twelve thousand years since the advent of agriculture,
we've squeezed about as much out of land farming as we can. For
humans to continue to thrive and expand across the Earth and

beyond, a new pillar of agriculture must be created. Within the last sixty years, the advent of industrial aquaculture (ocean farming) has provided the promise of such. Farming of the oceans, an area that covers more than 70 percent of the Earth's surface, will provide the solutions to our growing population's food problems and climate change, and repair our oceanic ecosystems. To advance our civilization into the future, we must return to the ocean ecosystems that cradled our infant species.

Aquaculture is the future of humanity. I believe this with such conviction that the day I was exposed to the industry of shellfish farming, I promptly quit my TV production job and completely rearranged my life to pursue a career in the shellfish farming industry.

For over a decade I worked as a freelance producer and cameraman on productions for the Discovery Channel, HGTV, National Geographic, NBC/Universal, and more. To produce a TV show, one must absorb into the subject matter. The better one understands the subject, the more informed choices one can make in telling the story and deciding which characters and settings to show, what information to convey, and how best to convey it.

Over the course of several months of research, I learned about the ways in which oysters can renew ocean environments, feed our growing human and animal populations, combat climate change, and sequester carbon. I filmed oyster farmers on location and I saw firsthand the sustainability of farming a food without the need for fresh water or nutrient inputs. I was hooked. I wanted to be an oyster farmer.

Producing TV shows no longer seemed fulfilling. I had glimpsed into the world of a profession that enabled local food production, bettered ocean biodiversity and environments, created local economic wealth, promoted science, sequestered carbon, and helped combat climate change. How could I resume my day-to-day existence when I knew there was something more contributing to society right around the corner? I decided to make a calculated career change.

My brother and I hatched a plan to start an oyster farm on Martha's Vineyard, an island off the coast of Massachusetts that we called home. I absorbed all there was to know about farming oysters. We acquired a boat and converted it for farming use. I mentored under one of the best farmers on Martha's Vineyard and earned firsthand working knowledge of ocean farm activities.

After a lot of growing pains, and a two-and-a-half-year application process, in 2014 my brother and I founded the Cottage City Oysters farm in Oak Bluffs, Martha's Vineyard. I was thirty-two years old, newly married, building a house, and had just given up my career to try something I had zero experience in.

Over a decade later, I've never regretted my decision. Shellfish farming not only has the potential to solve a number of the global challenges facing the current generation but also has a way of making the humans who work in the industry more connected to the environment, their communities, and our rich oceanic history—and more invested in the future.

The Cottage City Oysters farm has obtained international stature as a leader in the industry. We were the first open ocean shellfish farm in New England and the first in Massachusetts permitted to grow sugar kelp (edible seaweed). Our 3D regenerative farm model provides one of the most sustainable forms of protein farming on the planet, and scientific partnerships in conjunction with our operation have yielded multiple publications in scientific journals.

Our attention to detail, understanding of the local environment, and desire to create a high-quality oyster has made the Cottage City oyster one of the most expensive and sought-after oysters in the world. Our product is exclusive to Martha's Vineyard, EU Organic–certified, sculpted to perfection, and very clean. We strive to push the boundaries of the artisan craft of oyster farming.

The pages in this book have been fleshed out over the years from inspiring conversations with guests aboard the Cottage City Oysters farm tour. Every summer, thousands of curious oyster connoisseurs join us aboard *Leeward*, our thirty-foot tour vessel, for a one-hour BYOB voyage out to the oyster farm. The tour consists of the highlights of this book, followed by a lively, multifaceted conversation about oysters, the culture, and the future of humanity, all while drinking our favorite beverages, eating an endless supply of freshly harvested oysters, and cruising around in one of the most beautiful and pristine ocean environments the world has to offer. Some of these visitors are famous politicians, business professionals, celebrities, authors, and scientists. Most of them are everyday people like you and me. No matter how different the background of our guests, the collective consensus is that the vast majority know little to nothing about oyster history or culture. Through these conversations I've been able to identify some of the reoccurring general thoughts and curiosities about the industry, which are explained in detail in this book.

Through my journey, research, and hands-on education, I have become obsessed with everything "oysters." I have literally read every book on the subject. I think, act, and breathe a life that revolves around the husbandry of oysters. I aspire to know as much about the subject as anyone. My days begin with checking on the farm, boats, and gear. I research what other farms do and serve on multiple industry boards to help shape the future of the industry. My home is filled with oyster shirts, oyster-decorated coffee mugs, and an endless supply of oysters in the fridge. My front yard is full of oyster cages, bags, boats, and work rafts. My children are growing up in a house consumed by ocean critters, nautical lessons, and oyster knowledge. In all my time and energy, I have yet to find a book that tells the global history of oysters, highlights the various farm methods currently in use, and details how the industry works, where it's going, and why we need to expand it. This is the book I have been searching for, an all-encompassing journey of everything oysters—past, present, and future.

This book is an attempt to shed light on one of the unique histories of our planet: the humble oyster. Oyster farming has a distinctive global history, but it also has the potential to solve a large percentage of our food and climate change issues of today. With proper implementation of shellfish into our global food system, we can reduce the effects of climate change, feed our growing population, and restore the biodiversity and ecosystems we as humans have destroyed along the way. I hope this story educates about our past, presents solutions for our future, and inspires you to follow your own interests, whatever age you may be.

Dan Martino
Co-Owner | 50-ton Captain | Farmer
Cottage City Oysters / Martha's Vineyard Seaweed

SHELLFISH
DIFFERENCES

I T IS ESTIMATED THAT THERE ARE OVER TEN THOU-
sand different species of bivalve shellfish currently found living
in freshwater or saltwater environments on Earth. A bivalve is
defined as a shellfish organism having two outer shells that encase
and protect the animal's soft inner tissue. Organisms with only
one shell are defined as gastropods, which comprise snails, conchs,
octopuses, squids, and more than sixty thousand other identified
creatures on Earth.

Bivalves have existed on Earth for at least 510 million years
(since the middle of the Cambrian Period). The most common
bivalve shellfish are clams, mussels, oysters, and scallops, and while
these species have much in common, they have also evolved slight
differences along the way to enable the colonization of every possible niche within the marine environments, from the shallow inland

freshwater rivers to the ocean's deepest hydrothermal vents and everywhere in between.

The most common denominator of all bivalves is that they are filter feeders that pull water into their gills to breath oxygen and extract phytoplankton for food from the surrounding environment. When it comes to nutrition, bivalve shellfish are considered one of the healthiest, nutritionally dense foods available for humans. Bivalves build their shells from calcium carbonate, the same material as limestone, which helps sequester carbon from the atmosphere. From here, the differences between bivalves expand into an array of evolutionary tricks and specialties.

Dan Martino, *Shellfish Species; Mussel, Scallop, Oyster, Clam*

CLAMS

Clams come in a large variety of shapes and sizes but are typically rounded or oval. Clams call both freshwater and saltwater environments home, and some clams are of the pearl-producing kind. The oldest animal in the world is actually an ocean quahog clam, known as "Ming the Clam," who lived to be 507 years old, a Guinness World Record.

Clams have the ability to crawl or dig by utilizing a single foot. While most clams bury themselves in the sediment, a species of clam called *Calyptogena magnifica* was recently discovered to live solely around the edges of hydrothermal vents in the abyssal zone (thirteen thousand to twenty thousand feet deep), a region of the ocean which remains in perpetual darkness. Temperatures around these vents reach over seven hundred degrees Fahrenheit. Instead of filtering phytoplankton for food, the *magnifica* clam harbors a type of bacteria in its gills that oxidizes the hydrogen sulfide that seeps from the vents. The clam absorbs the nutrients produced by these bacteria.

Besides the *magnifica* clam, most clams find solace buried beneath the marine sediments. To dig, the clam first relaxes its shells, which allows them to open. The clam then extends its foot into the sediment to the furthest downward extent possible and pumps water into the foot to create a swollen kind of anchor. The clam contracts the shells, making it smaller in size, which allows water to surround and soften the sand around it. It then slides down onto its foot and the process is repeated until the clam reaches a maximum depth, sometimes as deep as eleven inches.

To breath and feed, the clam will extend a siphon into the waters above where oxygen and phytoplankton are "inhaled" and filtered. At low tide, the clam retracts the siphon and is hidden from predators beneath the sediment. Typically, a small indentation is formed in the sediment when the siphon is retracted, which is used by predators and fishermen to identify where clams are hiding. Fishermen harvest clams by raking the sandy bottom where these holes appear, digging the clams out of the sediment into their rake basket.

Clams are eaten raw or cooked around the world and celebrated as a tremendously healthy protein source. While clam gardens were used by Native Americans thousands of years ago to boost natural clam production, clam farming is considered a relatively new profession, having started in the 1960s. To farm clams, seed is typically purchased from a hatchery and planted onto leased tidal flats. A net is placed over the seed to protect it from predators. The clams grow to market size and are harvested by hand at low tides. Today, with diminished wild stocks, clam farming is in high demand and has increased greatly since the early 1990s.

SCALLOPS

THERE ARE ONLY two general types of scallops: the sea scallop and the much smaller bay scallop. Found only in saltwater environments, scallops inhabit the shallows of all oceans around the world and are considered the only "free-living" bivalves, capable of traveling some distance along the seafloor by clapping their shells together and expelling water out in a jet-like propulsion. The ability to flee is an evolved predatory response that is heightened by the numerous bright-blue eyes that surround the edges of the scallop shells. These simple eyes (sometimes numbering as many as two hundred) allow the scallop to see movement and predators by using telescope-like mirrors to catch light. Scientists believe the scallop's "brain" inputs the two hundred images and combines them to form one all-encompassing "vision" of the surrounding environment.

Scallops have saucer-shaped shells with fluted or ruffled edges and are found lying on the surface of the bottom sediment. Predominately a wild-caught species, scallops are typically harvested with a powerboat and dredge. With the decline of wild stocks, scallop aquaculture was invented in Japan during the 1960s. Farmed scallop production has been on a steady rise since, with about 90 percent of farmed production coming from Japan and China. Traditionally, just the adductor muscle of the scallop is eaten, and therefore scallops are not considered as nutritionally rich in vitamins and minerals as the other bivalves.

MUSSELS

MUSSELS ARE ONE of the oldest known sustenance foods for early humans, calling both freshwater and saltwater environments their home. Some of the most extreme mussels have been found living attached to the hydrothermal vents of the deep oceans, while pearl-producing mussels have been discovered inhabiting freshwater inland rivers.

Shells of the mussel are typically elongated and asymmetrical, with the "blue mussel" being the most popular variety eaten, with more protein and iron per calorie than a fillet of steak. Mussels use byssal threads (a type of string-like material) to attach themselves to rocks and each other, which historically allowed them to form enormous mussel beds along the rocky coasts of the world. Overharvest and poor water quality have driven the wild stocks to record low numbers, with freshwater mussels being the most endangered group of organisms in the United States. Luckily, mussels are one of the easiest shellfish species to farm and have a very quick seed-to-market turnaround (about six months to two years). The Food and Agriculture Organization (FAO) estimated 2018 global mussel production to be 2.1 million tonnes with a value of $4.5 billion.

OYSTERS

FOR MORE THAN 250 million years, the oyster anatomy has gone unchanged, a testament to the flawless design of nature, evolution, and the ability of an organism to exploit its local ecosystem.

Sharks, tardigrades, jellyfish, starfish, horseshoe crabs, nautiluses, and oysters are some of the few living "fossils" which share this remarkable trait, to seemingly defy evolution over hundreds of millions of years.

This teaches us that evolution can reach a state of perfection—a point in time at which the organism's design has become perfectly suited to exploit the niche within the environment it originally explored. Not only are these "end of evolution" designs flawless during a stable environment, but they are also capable of surviving the worst of times. Every animal listed above has survived at least three global mass extinctions and is currently living through the human-induced "Anthropocene extinction."

By understanding the oyster anatomy and function, we can come to understand the ideal environment that the oyster flourishes in, which in turn allows us to grasp the ways in which we can augment the environment to accent various attributes within the oyster.

No book better illustrates and describes the anatomy or function of the oyster than *The Oyster*, written by William K. Brooks and originally published in 1891. Accompanied by beautiful hand-drawn illustrations by A. Hoen & Co., the book accurately described for the first time how the oyster anatomy looked and functioned. Brooks's narration reads like the journal of an obsessed scientist, discovering in real time where each part fit, and rightly so—he literally wrote the book on the subject.

While much more has been learned about oyster anatomy since Brooks's account, *The Oyster* still remains the starting point on the subject.

According to Brooks,

The most prominent fact in the organization of the oyster is its shell. Its body is shut in between two long concave stony doors, which are made of limestone, and are fastened together at one end, somewhat in the same way that the covers of a long, narrow check-book are bound together at the back. One of these shells, the flat one, is on the right side of the body, and the other, which is much deeper, is on the left. When oysters are fastened to each other or to rocks, the left shell is attached, and the oyster lies on its left side. When it is at home and undisturbed its shell is open, so that the water circulates within it, but when disturbed it shuts its shell with a snap, and is able to keep it firmly closed for a long time. The snapping drives out the water, together with any irritating substances which may find their way in, and on the natural beds the oysters snap their shells shut, from time to time, for this purpose. *

Lining the inside of the shell is the *mantle*, and the edges of the mantle are lobed into three folds. The outer fold secretes the shell, the middle fold contains dark fringed, hair-like *sensory tentacles,* and the inner fold is muscular but also has tentacles.

The tentacles are the oyster's connection to the outer world and are partially exposed when the shell is open. As the oyster grows, it first sends the muscular tentacles out like structural probes, "feeling" for environmental objects to bind to or shape around. Using the outer fold, the oyster then secretes a thin prismatic layer of shell across its entire inner shell, starting near the *hinge* and ending at the

---

* William K. Brooks, *The Oyster: A Popular Summary of a Scientific Study* (Baltimore: Johns Hopkins, 1891), 15.

extending tentacle probes. The shell cascades in between the tentacle probes, and after the new layer has hardened, the process starts over, with each new layer of shell slowly thickening the whole. The hinge is always thickest, being the oldest and first shell created by the oyster.

If the sensory tentacles sense danger, they send a signal through the oyster's ganglion of nerves, which strike the *adductor muscle*, causing it to contract and in turn close the shells tightly shut.

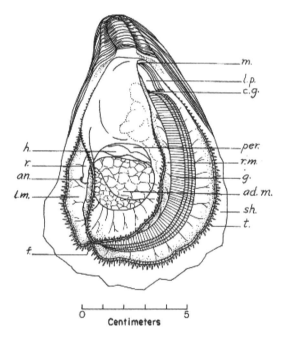

Organs of *Crassostrea virginica* as drawn from a live specimen. ad.m.–adductor muscle; an.–anus; c.g.–cerebral ganglion; f.–fusion of two mantle lobes and gills; g.–gills; h.–heart; l.m.–left mantle; l.p.–labial palps; m.–mouth; per.–pericardium; r.–rectum; r.m.–right mantle; sh.–shell; t.–tentacles.
Source: P. S. Galtsoff, "The American Oyster, *Crassostrea virginica* Gmelin"

The adductor muscle is actually constructed of two distinct types of fibers, called the *quick muscle* and the *catch muscle*. The quick muscle is the first to respond to danger, able to quickly seal the shells shut, but unable to hold prolonged contraction. This is where the catch muscle fibers kick in, reacting more slowly but able to withstand long periods of contraction with ease. At rest, the oyster shells remain open, held apart by a rubber-like ligament in the hinge.

The *visceral mass* fills the majority of space between the hinge and adductor muscle and contains a large fold that creates the right and left mantle skirts. Together they enclose a large water space, the *mantle cavity*, which is divided into *inhalant* and *exhalant chambers* by the *gills*.

Attached to the gills are hair-like structures called *cilia*, which move like pieces of seaweed wafting in the ocean current; however, it is their movement that creates the current! As these tiny tentacles wave, they generate a current that sucks water into the oyster's inhalant chamber.

As many as fifty gallons of water a day are pulled into the inhalant chamber by the cilia. The water passes into the pores, or *ostia*, in the walls of the gills, and once through, the now-filtered water enters the exhalant chamber and *cloaca*, where it is discharged back into the environment.

The oyster gills are one of the greatest marvels of evolution; a matrix of folds, nooks, and crannies with the ability to capture micron-sized specks of material present in the water. As these captured materials consolidate within the gills, they are wrapped in a mucous string and transported by cilia's water current to the *labial palps* located just outside the *mouth* for sorting.

The labial palps are able to sort through the various specimens of phytoplankton, seeking out specific strains depending on the oyster's needs. After sorting is complete, the cilia transport the required phytoplankton into the oyster's mouth to be swallowed and sent to the oyster's *stomach*, where it is broken down in the *ceca* by enzymes, intracellular digestion, and absorption.

Bits of grit and detritus that cannot be used as food are wrapped in mucus and expelled as pseudofeces, a rich fertilizer utilized by ocean bacteria and seaweed for growth. Digested phytoplankton exits the oyster through the *anus* in the form of pellets, another rich fertilizer ocean bacteria use for growth.

Inside the visceral mass is the pericardial cavity. Within this cavity is the oyster's *heart*, which consists of a ventricle and two atria. The atria drain oxygen from the gills and send it to the ventricle to be pumped throughout the oyster via arteries.

The nervous system of the oyster is simple, very weakly developed, and small, consisting of a ganglion of nerves connecting from the mantle sensory tentacles to the adductor muscle.

While the finer details of anatomy are more technical than most wish to comprehend, the most important takeaway is that the oyster is like a fine-tuned machine with the sole purpose of filtering and extracting the most minute elements out of seawater—from the single cells of phytoplankton to the microscopic free-floating amino acids. Besides eating, the only other activity for oysters to do is reproduce.

The reproductive system of the oyster consists of the *gonad*, which occupies most of the space in the visceral mass between the ceca and the stomach. The gonad is an indistinct organ without definite walls and can be of either male or female sex, constantly changing year to year, possibly depending on the needs of the reef system. Depending on the sex, the gonad will fill with either sperm or eggs, which will release into the environment for fertilization.

Millions of eggs and sperm are released by an individual oyster; however, the large majority of them never fertilize, instead finding life at sea a risky endeavor, ending as a snack for hungry ocean organisms.

At the end of Brooks's account, he argues for the complete human intervention of oyster fertilization: "A thorough knowledge of the oyster . . . will show the capacity of the oyster for cultivation, and it will also show why its cultivation is necessary, and why our resources can never be fully developed by oysters in a state of nature."

Brooks explains how hatchery cultivation can be much more successful than natural oyster propagation and illustrates the amount of mass that can be acquired through these processes:

An average Maryland oyster of good size lays about sixteen million eggs, and if half of these were to develop into female oysters, we should have, from a single female, eight million female descendants in the first generation, and in the second, eight million times eight million or 64,000,000,000,000. In the third generation we should have eight million times this or 512,000,000,000,000,000,000. In the fourth, 4,096,000,0 00,000,000,000,000,000,000. In the fifth, 33,000,000,000, 000,000,000,000,000,000,000 female oysters, and as many males, or, in all, 66,000,000,000,000,000,000,000,0 00,000,000,000. Now, if each oyster fill[s] eight cubic inches of space, it would take 8,000,000,000,000,000,000,000,0 00,000,000,000 to make a mass as large as the earth, and the fifth generation of descendants from a single female oyster would make more than eight worlds, even if each female laid only one brood of eggs.*

Brooks is correct in his assumption that human intervention can greatly increase oyster fertilization. We now create 98 percent

---

* Brooks, 44–45.

fertilization success rates of oyster eggs within hatchery-controlled environments. With these kinds of numbers, just a handful of oysters can supply enough spat for every farm on the entire east coast of America, every year.

The rest of William K. Brooks's book makes the argument for oyster aquaculture on a global scale, as quickly as possible.

If only the masses had listened.

Soon after *The Oyster* was published, global industrialization and overharvest wiped out the last remaining oyster beds around the world. It would take almost one hundred years after the publishing of his book for developed countries like France and America to begin oyster aquaculture.

Today, the global oyster population is around just 1 percent of their historic numbers.

THE PAST

IN THE BEGINNING

ROUGHLY 4.54 BILLION YEARS AGO THE EARTH was formed. Evidence now tells us that a vast amount of water was trapped within the original building blocks of Earth, so the oceans formed as the Earth cooled, around 4.5 billion years ago. For almost a billion years afterward, the Earth was bombarded by asteroids and comets, rendering the surface inhabitable to life. As the Late Heavy Bombardment period subsided, life immediately sprang forth some 3.7 billion years ago, as evidenced by fossils. We call this most primitive form of life Cyanobacteria, a form of phytoplankton that obtain their energy through photosynthesis. Cyanobacteria are from the family of organisms called prokaryotes and are frequently referred to as blue-green algae, which is a common mistake since algae are considered eukaryotes.

Using the sun's energy and the nutrients present in the rich primordial ocean waters, Cyanobacteria grew like a type of plant or algae; they inhaled carbon dioxide and exhaled oxygen, just as they still do today. Cyanobacteria are the first organisms known to have produced oxygen on Earth. When the Cyanobacteria died, they sank into the ocean sediment, locking away the fragments of carbon that they had stored during life. In this way, Cyanobacteria have terraformed our planet into an oxygen-rich atmosphere and have sequestered carbon for billions of years.

The Cyanobacteria's offspring multiplied, slowly at first, and then became unstoppable, compounding mass upon mass until they circumvented the Earth. For about four billion years, life was limited to this simple, single-celled organism design. As the phytoplankton evolved, so too did its motivation. The original single-cell organisms lived harmoniously with the world, consuming sunlight energy and carbon while exhaling oxygen, but then in a twist of fate, some of the organisms decided they would rather consume their neighbor for energy than capture it from the sun themselves, and so began the predator and prey relationship. From this moment on, life branched into two distinct trees: organisms that gather their energy from the sun, and organisms that gather their energy from other organisms. In response to the predators, the prey organisms grew larger and multicelled, for predators cannot eat what does not fit into their mouths. We see this evolutionary tactic repeated throughout all of history, from the largest trees and dinosaurs to the giant whales of modern day. As the predator/prey competition evolved, the designs became more complex and eventually led to the creation of other competing bacteria, phytoplankton, sponges, mushrooms, starfish, and corals. This early period of life, from 3.7 billion to about 540 million years ago is referred to as the Precambrian.

After four billion years of Precambrian life inhaling carbon, exhaling oxygen, dying, and sequestering the atmospheric carbon into the sediment where it still lies buried today, the Earth's atmosphere became rich with oxygen. A pivotal point in history came about 540 million years ago when this excess of oxygen in the atmosphere caused what is known as the Cambrian explosion—an event that occurred over a period of about eleven million years and allowed evolution to take a giant leap forward. It was an "explosion" of various forms of life unlike anything the Earth had seen before. For the first time, organisms became complex.

Life was still confined only to the ocean shallows during the Cambrian explosion; however, for the first time hard body parts, such as shells and backbones, began to evolve. These early shellfish contained just one shell, which was used for protection from the ancient predators like the carnivore Anomalocaris. The shellfish also learned to swim or dig themselves into the ocean sediment for protection. They were filter feeders, filtering ocean water for the rich nutrients and algae that it contained. They are the earliest descendants of all living shellfish today, and some of the oldest forms of life on Earth.

Over the next 250 million years, animals began to inhabit the land, and shellfish continued to evolve, branching off and forming many different specialized types. Some evolved to grow two shells, becoming bivalves. This is the first know appearance of oysters. According to fossils, these early oysters could grow as large as three feet and weigh up to twenty pounds. Before long, the ancient oysters quickly spread around the globe and filled every filter-feeding niche available in the oceans.

About 250 million years ago the Permian Period ended with the greatest mass extinction in Earth's history. Ninety percent of all living species on the planet went extinct, most likely due to a buildup of methane in the atmosphere and a superheated greenhouse gas effect. The oceans became highly acidic. Oysters and other shellfish were hit hard by this event but were able to quickly evolve, survive, and reestablish themselves around the world.

Shellfish thrived in the years following the mass extinction. What doesn't kill you makes you stronger.

The oldest fossil oysters (pre–Permian extinction) built their shells entirely out of aragonite (a hard carbonate material), but later changed to a calcite-based shell like their modern counterparts after surviving the extinction event. This change in shell structure was most likely an evolutionary response to ocean acidification. As a result, we see numerous oysters in the fossil record surviving this period due to the more indestructible material of their carbonite shells.

The fossil record dates relatives to the modern-day oyster as far back as the Triassic Period, more than 200 million years ago, in what is now far eastern Siberia and Ellesmere Island in Arctic Canada. The climate of this time was warm and dry, with little to no ice at the polar regions. The giant supercontinent Pangea was beginning to break apart into the smaller continents we know today, and dinosaurs roamed the lands in search of food. Imagine hungry dinosaurs looking for a meal. Starved, they stop along the ocean bank for a quick snack of a dozen oysters. Who could blame them?

Interestingly, modern newborn oysters still begin their lives by building their shells with aragonite, like their ancient predecessors. Only after the oyster goes through a metamorphosis does it change from making aragonite shells to making calcium carbonate shells, mimicking the history of the oyster's evolution.

This is a beautiful example of recapitulation theory, also called the biogenetic law or embryological parallelism—often conveyed via zoologist Ernst Haeckel's phrase "ontogeny recapitulates phylogeny."

This theory suggests that the development of the embryo of an animal, from fertilization to gestation or hatching (ontogeny), goes through stages mimicking the evolution of the animal's remote ancestors (phylogeny).

Amazing, right? Our fetal development growth mirrors the stages of our evolutionary path. This natural phenomenon can be found throughout the animal kingdom, including human fetal development. The most common example is a frog, which first begins life by mimicking its fish evolution—the tadpole, before it sheds the tail and gills for legs and lungs.

After the Precambrian extinction and throughout the following Triassic Period, oysters quickly dominated the ancient oceans, cementing themselves as a "pillar" species of the environment and ecosystems around them. Where coral reefs could not form, oyster reefs took their place. Like coral reefs, oyster reefs have the ability to change the marine environment around them, provide shelter to a plethora of marine organisms, and provide nutrients for bacterial and seaweed growth. Oyster reefs "aquaform" the surrounding ocean environment (like terraforming, they create a set of parameters for all other life forms to exist within).

As the Earth became more stable, many branches of oysters evolved, each equipped and better suited for the microclimates around the world. In response to oyster global domination, an array of more complex animals developed, each reliant on the ecosystem created by the oysters.

The immensity of the historical oyster population cannot be overestimated. The amount of oyster biomass that has been created on Earth is most likely the largest of any species in history. Many of the modern-day hills and mountains are the remains of these ancient oyster reefs, which now exist far above sea level due to "modern" uplifting of the continents. If the Earth were named after the most dominate species to have ever existed, it would be called "Planet Oyster." Oysters are one of the oldest-surviving, most-successful species of life on Earth.

For more than four hundred million years, trillions upon

trillions upon trillions of oysters have evolved and survived multiple extinctions, ice ages, asteroid impacts, dinosaurs, and acidic oceans. Today more than two hundred different species of oysters cover the Earth, but it should come as no surprise that the one thing that has brought this incredibly successful species to the brink of extinction is humans. Humans have now consumed approximately 99 percent of the historical wild population of oysters. In essence, we have destroyed one of the largest and most important building blocks of the ocean ecosystem, without ever knowing the balance it provided.

WHEN OYSTERS MET
HOMO SAPIENS

OMO SAPIENS EVOLVED IN AFRICA SOMETIME between 300,000 and 200,000 years ago, according to anthropologists' assessment of fossil, genetic, and archaeological evidence. Then, around 195,000 years ago, the global climate entered a period of cold and dry conditions that lasted around 70,000 years, a phase referred to as Marine Isotope Stage (MIS) 6.

Fossil records show that during the MIS 6 episode, the *Homo sapiens* population plummeted. At this time, all of Europe was buried beneath a giant sheet of ice. The interior of Africa was under such severe drought that much of the continent would have been uninhabitable. Modern genetic DNA testing tells us that the human population declined from near ten thousand breeding adults to maybe as few as just six hundred.

Everyone alive today is descended from this small group of human survivors who migrated to a single region off the southern coast of Africa. Known as Pinnacle Point, a small rocky bluff immediately south of Mossel Bay, South Africa, this region is believed to have been the last holdout for *Homo sapiens* to survive the global catastrophe. Sandwiched between the coastline and a lush grassland ecosystem, Pinnacle Point provided a productive, predictable, rich refuge zone for early modern humans.

Excavations of the Pinnacle Point area began around the year 2000 on a series of caves, where paleoanthropologists found evidence that Middle Stone Age people occupied the caves between 170,000 and 40,000 years ago and were systematically using ocean resources (shellfish) for survival—the earliest evidence of humans having done so. In essence, our surviving population turned to the most stable and historically significant food source that they knew of to survive: shellfish.

Shellfish and seaweeds have been consumed by humans since the dawn of our species. Our survival and evolution are so intertwined with shellfish that many theories of human evolution revolve around the human-shellfish relationship. For instance, some scientists postulate that our brain growth (the expansion of self-awareness and consciousness) was due to an excess of omega-3 fats ingested from a stable shellfish diet. Evidence of this relationship can be found in the Pinnacle Point PP13B cave, where the first human-made pigment dyes used for ceremonial (religious) practices can be found next to piles of discarded shellfish meals.

The gathering of shellfish from intertidal zones required early humans to understand the relationships of the ocean tides, the lunar cycles, and the local weather patterns, making early *Homo sapiens*' survival dependent on seasonal observation and timekeeping. Perhaps this is how the first concept of time was created, when it was recognized that the high tide would fall to a low tide every twelve hours, and that twelve hours past low tide the high tide would return again. It seems plausible that if their meal depended on a tidal swing, they would surely count the interval in between meals. Perhaps this is also when a numbering system of some sort was developed, to count the time between tides and meals.

Other theories suggest that *Homo sapiens'* expansion out of Asia, across the land bridge into North America, then down the West Coast into South America was due to early humans following seaweed and shellfish (the Kelp Highway), a familiar source of food. This is the only plausible reason for the much earlier development of South America and the much later development of the East Coast of North America.

Regardless of the historical period or geographic location, shellfish and specifically the oyster has been a staple food for every culture of humanity that has had access to the species. The oyster as a food source is perhaps the *most* unifying factor between human cultures over time. Throughout human history, the oyster has been a constant character and has carried a lore of luxury, health, refinement, and success. Oysters have honorable mention in many classic songs and stories and appear in paintings to signify wealth and abundance.

The earliest mention of oysters in literature comes in the sixteenth book of Homer's *Iliad*, written sometime around the late eighth or early seventh century BC. In the ancient Greek story, Patroclus, Achilles's comrade-in-arms, mocks Cebriones, Hector's charioteer, when he falls dead to the ground after having his forehead crushed by a jagged rock that Patroclus had thrown.

Homer writes:

"Hah, look you, verily nimble is the man; how lightly he diveth! In sooth if he were on the teeming deep, this man would satisfy many by seeking for oysters, leaping from his

ship were the sea never so stormy, seeing that now on the plain he diveth lightly from his car. Verily among the Trojans too there be men that dive."

Homer's imagery of a man jumping into the depths of a rough sea to fetch oysters for others to enjoy must have been based on personal or anecdotal accounts, for it clearly demonstrates the knowledge of deep-sea oyster reefs existing within the Mediterranean and the common method of ship diving for their harvest.

The gathering and consumption of wild oysters is as old as humanity; however, the cultivation or farming of sea animals is not nearly as old as agriculture (thought to have been created around 12,000 BC).

While not focused on oysters specifically, the Chinese are credited with the first origins of aquaculture (farming marine organisms). A 2019 team of German, Japanese, Chinese, and U.K. scientists uncovered evidence at a Neolithic settlement called Jiahu that showed a group of Chinese Stone Age people farming common carp, a freshwater fish still present today, as far back as 6200 BC to 5700 BC.

There is even some speculation that the Chinese at Jiahu began cultivating rice paddy fields around this same time and that they may have placed the carp into the paddy fields to create a type of agriculture-aquaculture farm system, in which the carp ate the weeds and insect pests that fed on the rice, and in return the carp helped to fertilize the rice paddies. This practice has existed since at least 1200 BC and is still very common today. While the origin of this symbiotic relationship is still not proven, the evidence that the settlement was able to control water channels and artificial

---

* Homer, *The Iliad*, bk. 16, line 745

ponds to some extent could mean that both rice cultivation and carp aquaculture developed in tandem. A cutting-edge Neolithic town, Jiahu is also one of the first places where inhabitants learned to make wine from fermented honey and rice, build musical instruments, and carve symbols, which may be one of the earliest-known predecessors of writing.

Older than Stonehenge and the Egyptian pyramids, another ancient culture known as the Gunditjmara people from southwest Victoria (Australia) began cultivating ocean critters around 4000 BC. Ruins support evidence of six-thousand-year-old networks of eel traps in Lake Condah and Darlot Creek, demonstrating some of the oldest surviving evidence of aquaculture. The eel farms cover over seventy-five square kilometers and consist of artificial channels and ponds for separating eels, as well as smoking trees that were used to preserve the eels for export to other regions of Australia.

The traps consisted of a series of canals and graded ponds where the Gunditjmara people manipulated the water levels to encourage eels to swim into holding ponds and then placed funnel-shaped baskets at the spillways between ponds to capture large eels for harvest. Smaller eels are thought to have been allowed to slip through, ensuring future generations' survival. Eels were transferred from pond to pond and allowed to grow until they reached edible size. This form of aquaculture allowed the Gunditjmara people to settle in one central place as opposed to following a nomadic lifestyle, and accordingly, they built stone houses around the lake to support a large community.

Around 1500 BC the Egyptians had begun placing tilapia, a common fish found in the Nile, into artificially made captivity ponds, a type of aquaculture similar to the ancient Chinese carp method. Known as "Assyrian vivarium," these ancient household versions of a lobster holding tank could be found in wealthy ancient

Assyrian and Roman houses, stocked full of fish and crustaceans.

Greek philosophers Aristotle (384–322 BC) and Xenocrates (396–314 BC) "noted the cultural preference and value of oysters [and] highlighted different harvest strategies, including transferring oysters from areas of abundance to locations where they grew faster and tasted better."* While this does not constitute "farming," it does demonstrate the understanding of merroir (the different flavors of oysters influenced by location). These early experiments of transferring oysters to better waters undoubtably led to oyster cultivation.

Another ancient oyster treasure is hidden in the Louvre, where there is a papyrus dated to the second or third century BC containing a riddle in the form of a six-line poem. Transcribed by Peter John Parsons, British classicist and academic specializing in papyrology, the riddle reads:

> *Near the earth tomb of Ethiopian Memnon, (it was not the Nile which reared me, it was Ocean which) nursed me on the rocks of adamant (the virgin, Helle?), as I revelled in the sweet rays of Agrotera (Artemis, the moon). I am a feast without fire (uncooked) for mortal men, when Doso's bridegroom (Aphrodite's lover, Ares, the knife) cleaves me with his hide-piercing weapon. What am I?†*

---

* Robert Botta, Frank Asche, J. Scott Borsum, and Edward V. Camp, "A Review of Global Oyster Aquaculture Production and Consumption," *Marine Policy* 117 (July 2020), doi.org/10.1016/j.marpol.2020.103952.

† P. J. Parsons, "The Oyster." *Zeitschrift Für Papyrologie Und Epigraphik* 24 (1977): 1–12.

The answer to this ancient riddle—an oyster.

The word "ostracize" has ancient Greek origins associated with the Greek root word *ostreion* for "oyster." In ancient Greece, ostracism was the name of a legal political practice by which men deemed dangerous to the liberties of the people were banished for ten years by public vote. To assemble the votes (sometimes as many as six thousand) for tally, the accused names would be written on the side of an oyster shell, one of the most abundant objects available in ancient Athens. Once all the shells were collected, the tally would be counted. What a delicious way to vote.

Aphrodite, the Greek goddess of beauty and love, was said to have emerged from the sea on an oyster shell. She then gave birth to Eros, the god of intimate, erotic, and romantic love. The word "aphrodisiacs" comes from the goddess herself.

Casanova, the legendary Venetian lover, is said to have eaten fifty raw oysters every morning for breakfast, typically warm and off a woman's breast. Casanova believed deeply in their stimulating effects and helped to bolster the oyster as an aphrodisiac.

Cleopatra, the famous Egyptian queen, is said to have boasted about her extravagantly lavish parties at which thousands of oysters were consumed. In a wager between Mark Antony and herself as to who could throw the most expensive party, Cleopatra solidified the win by crushing up a pearl from her earring into a cup of vinegar and drank the mixture once the pearl had dissolved. The pearl was undoubtedly sourced from an oyster.

Describing the event, Roman naturalist and philosopher Pliny the Elder (23–79 AD) wrote in his book *Natural History*,

There have been two pearls that were the largest in the whole of history; and both were owned by Cleopatra, the last of the Queens of Egypt—they had come down to her

*through the hands of the Kings of the East. When Antony was gorging daily at decadent banquets, she, with a pride at once lofty and insolent, a wanton queen, poured contempt on all his pomp and splendor, and when he asked what additional magnificence could be contrived, replied that she would spend ten million sesterces on a single banquet [more than $5 million today]. Antony was eager to learn how it could be done, although he thought it was impossible. Consequently, bets were made, and on the next day, when the matter was to be decided, she set before Antony a banquet that was indeed splendid, so that the day might not be wasted, but of the kind served every day—Antony laughing and expostulating at its cheapness. But she vowed it was a gratuity, and that the banquet would round off the account and her own dinner alone would cost ten million sesterces, and she ordered the second course to be served. In accordance with previous instructions, the servants placed in front of her only a single vessel containing vinegar, the sharpness and power of which disintegrates and decays pearls. She was at the moment wearing in her ears that remarkable and truly unique work of nature. And so, as Antony was full of curiosity to see what in the world she was going to do, she took one earring off and dropped the pearl in the vinegar, and when it was wasted away swallowed it. Lucius Plancus, the judge of the wager, placed his hand on the other pearl when she was preparing to destroy it also in a similar way, and declared that Antony had lost the battle—an ominous remark that came true.**

---

* Prudence Jones, "Cleopatra's Cocktail," *The Classical World* 103, no. 2 (Winter 2010): 207–220.

The infamous ancient spice trade route was also home to oysters. Around 300 BC, boatloads of spice and incense from ancient Indonesia, Sri Lanka, and India would land on the shores of the Persian Gulf, where a caravan of camels would await to bring the goods across a forty-six-day trek through the desert to the Mediterranean coast. Once the goods were sold, the empty camel caravan would load up on the most prized Mediterranean goods for the return trip. Olives, dates, wines, dried fish, and oysters were among the items taken for the trek home, as evidenced from archaeological digs.

Oyster culture was written about in the Chinese Han dynasty (206 BC to 220 AD) in the ancient book *Treatise on Pisciculture*, written by the famous Taoist and economist Fan Li. It was considered the first book written about aquaculture practices, and the basic principles described in *Pisciculture* went on to dominate two thousand years of Chinese fish farming practices.

While the consumption and transfer of wild oysters into better growing conditions dates back much further, the farming of shellfish, and specifically the oyster, is typically credited to the ancient Roman merchant, inventor, and hydraulic engineer, Sergius Orata around 100 BC.

Prior to 100 BC, wild oyster beds were fished by simply wading out into the waters at low tide and breaking bundles of oysters off the reef for consumption. Ancient Egyptians, Greeks, and Romans significantly depleted the wild oyster beds of the Mediterranean by 100 BC. As the oysters became harder to acquire, their value and demand continued to increase.

Where there is a need, there is an opportunity. Sergius Orata wanted to take advantage of the growing desire for oysters among wealthy Roman aristocrats and devised a new technique for breeding and growing large numbers of oysters in captivity. This was the first recorded instance of shellfish cultivation in history.

Orata's plan was simple: create a series of channels and dams from the ocean into nearby lakes, throw adult oysters in the lake, and wait for them to spawn baby oysters. Orata planned to capture those babies, grow them to a larger size, and sell them for top dollar to the aristocrats in the area.

There is question as to exactly which body of water he chose for his experiment. Most texts label Lucrine Lake as the main body of water; however, late texts also describe Lake Avernus as an oyster park, which contained oyster farms until recently. Lake Fusaro was likely also a part of the production, as their oysters were known throughout the ancient Roman world for their quality. Most likely all three bodies of water were used, due to their close proximity to each other and, per historical reference, their interconnectedness with channels and dams. All three bodies of water lay in the Campania region of Italy, just west of Naples and the ancient resort town of Baiae.

Regardless of which body of water he used first, Orata undoubtedly picked them because of their ideal cultivation characteristics. The lakes were small bodies of water separated from the ocean by thin sandbars. The small size of the lakes would have allowed them to heat up more quickly than the surrounding ocean waters, a necessity for spawning oysters. It is also likely that historically, oysters may have naturally grown in the lakes. However, by the time of Orata's experiment the wild oysters would have been consumed by the ancient Greeks and Egyptians long before him. The proximity of the lakes to the neighboring port of Naples and the resort town of Baiae made for easy transport to high-dollar markets for his finished farmed oysters.

The location and size of the salt lakes were near-perfect; however, the bottom sediment type was not. All three bodies of water are described in ancient texts as being void of all rocks or hard bottom structure. This is an important fact to note because baby oyster embryos require a hard substrate to attach to before they can meta- morphosize into an oyster.

Either through trial and error or superb observation, Sergius Orata came to understand how oysters spawned, what oyster babies needed for successful propagation, and how to raise baby oysters to a large enough size for market sale.

WHERE DO BABIES
COME FROM?

THERE ARE MANY TYPES OF EDIBLE FARMED oysters in the world and even more nonedible kinds. Most edible species today originated in Japan and were exported around the world to help restore diminished reefs, such as the Kumamoto and Pacific varieties. Europe has the native Belon oyster. North America has the native Olympia (West Coast) and Virginica (East Coast) oysters. Australia has the Sydney rock oyster. Regardless of the species, for the last 250 million years oysters around the world have made babies in pretty much the same fashion.

Oysters spawn during the summer months, months without an *r* (May, June, July, August). This is one of the reasons for the old expression, "Don't eat oysters in months without an *r*" The expression was most likely created by the Romans and served as an attempt to leave the oysters alone during the spawning months.

These months are also the hottest of the year (in the northern hemisphere) and when the rule was created, refrigeration had yet to be invented. A hot oyster was more likely to get you sick. Now, with modern hatcheries and refrigeration, this rule no longer applies.

Oysters do not have brains, do not have a central nervous system, and are not considered sentient beings, making them an exception for most vegans and vegetarians. Instead, they have what is called a ganglion of nerves, which permeates their flesh and shells and allows them to make binary decisions in reaction to their environment. What kind of decisions do oysters make? To filter seawater or not to filter seawater, that is the question.

Oysters also have the ability to change gender. Typically, the older they become the more likely they are to become female. There is a hypothesis that oysters are able to somehow sense what gender is needed within the reef system for improved propagation; however, this is yet to be proven. My wife says that they become female with age because they are wiser. This is also yet to be proven.

Oysters reach sexual maturity between one and three years of age, and their cue to spawn is decided by the surrounding water temperature. As the water reaches about seventy degrees for a period of two to three weeks, it conditions the oysters into spawning mode.

The males shoot out hundreds of millions of sperm. The females shoot out hundreds of millions of eggs. In the rare event that the sperm and eggs meet up while floating in the water column, an embryo is formed within forty-eight hours that has an eye spot, a flipper, and a foot. The eye spot can sense in black and white, which tells the larva which way is up (the sun), and which way is down. The flipper allows the larva to swim vertically in the water column. After about two weeks of skimming the surface waters for food, the larva begins to dive to the bottom of the seabed in search of a suitable landing site.

It is believed that the oyster larvae are able to sense amino acids that are filtered by adult oysters, and in this capacity they are able to seek out existing oyster reefs as a landing spot. If no reef is found, the larva will place its single foot onto the first hard substrate that it can find. This is typically a rock, a dock piling, or ideally another oyster. The larva then secretes a concrete-like substance around the foot and will stay attached to this spot for the rest of its life. After cementing itself, the larva's eye spot and flipper dissolve away, and it goes through a metamorphosis and becomes an oyster. Once attached to a substance, the oyster is now considered "spat."

It is important to note that the native European Belon oyster differs in this reproduction process in one very small detail—one that explains why Romans would not eat them in months without an *r*. The European female oyster does not shoot out her eggs but instead holds them in her gills while ingesting sperm from the water. In this way, the eggs can fertilize and grow in the safety of the mother's shell. If you eat a pregnant European oyster it will be full of small, sand-like crunchy babies—not exactly a desirable trait. Once able to swim, the European oyster embryos leave the safety of their mother and find a hard substrate to settle on.

For the rest of its life (up to twenty-five or thirty years), the oyster will stay cemented to the original hard substrate it attached to after birth. After metamorphosis, the oyster spat (about the size of a grain of pepper) begins to filter the water. Day after day, the oyster will filter the water, absorbing free amino acids and converting phytoplankton into rich minerals and protein. Dissolved calcium and carbon are converted into shell. The calcium carbonate shell will eventually sequester about twelve grams of carbon per oyster, helping to combat climate change. These shells can last over 150,000 years on land and for this reason, oysters are being considered for carbon credit trading programs.

Every two to three weeks the baby oyster will double in size until it reaches about one inch. This growth rate varies greatly with water temperature, species type, and the amount of food (phytoplankton) present in the water. More about this later.

If the ocean waters remain warm and phytoplankton is present year-round, the oysters will continue to grow and spawn. If the waters get very cold and the phytoplankton disappears, as happens in New England and other northern regions, the oysters will hibernate until the food comes back in the spring.

Oysters are the ultimate ocean-filtering machines. An adult oyster can filter up to fifty gallons of water a day! It is now understood that oysters will consume specific types of phytoplankton from their filters based on their nutritional needs at the time, which can vary throughout the season depending on site and climate.

If a grain of sand or other foreign material enters the oyster, it will spit it out. If the oyster is of the nonedible pearl-making kind, it will save the grain of sand, cover it with shell, and turn it into a pearl. These oysters are farmed exclusively for their pearl-making abilities and are not eaten.

The oyster castings (poop) are little balls of digested algae that embed into the soil and create fertilizer for bacteria and seaweeds. Similar to earthworm castings, oyster poop helps to rebuild soil and nourish thousands of microscopic organisms.

Oyster reefs are the coral reefs of the colder ocean environments and create the foundation for ocean life. Without oysters, the oceans would become polluted with algae, coastal erosion would be exacerbated, and juvenile sea creatures would have little habitat to grow up in. This is exactly what has happened to our oceans as we have harvested these animals to near extinction.

CULTIVATING OYSTERS

ROUND 100 BC, Sergius Orata discovered the magical process of oyster reproduction and he also noticed the fault within his lakes—the bottom sediment consisted of a soft, silt-like material that the baby oysters could not attach to. Orata used this to his advantage.

Without the hard surfaces, the baby oysters would suffocate in the lake bottom muck. To control the locations of baby oyster settling, Orata stacked the adult spawning oysters onto pyramids of submerged rock inside the lake. He tied bundles of submerged twigs around the mature oysters and waited. The oysters spawned and the embryos, having no hard substrate to attach to, were forced to attach to the floating twigs and surrounding rock pyramids. As the baby oysters grew larger, Orata would snap them off the twigs, place them into various grow-out ponds, and allow them to grow to market size. He could then scale and repeat the process infinitely, and he did.

At the pinnacle of his operation, Orata had all three lakes in production, as well as hand-dug canals and smaller ponds. He made a fortune with his operation and continued to improve his artificial beds, eventually lining the large lakes with clay roof tiles. The clay tiles provided the perfect hard substrate for the oyster embryos to attach to. In this way, he was able to mass produce oysters for the growing demand of the Roman upper class.

Sergius Orata's name became synonymous with luxury and wealth. His Lucrine oysters, as they were called, became known as the "Saddle-Rocks" of Rome, for their ease of carrying while riding on a horse—the original to-go cup. Orata's success with oyster cultivation was so widely known that the Romans had a saying that, if the oysters ever stopped growing in Lucrine Lake, Orata would make them grow on the houses' clay roof tiles.

Orata took his oyster empire wealth and went on to become a luxury villa real estate builder and the inventor of heated floors and hanging baths. Aside from oyster cultivation, his most long-standing contribution, however, came in the form of a lawsuit.

In a private-law case, a Roman tax farmer named Considius decided to sue Orata for using a public resource (the lakes) for private gain. The lawsuit was brought before the famous Roman orator and statesman Lucius Licinius Crassus, who ultimately sided with Orata. The outcome of this lawsuit has followed society into modern times. Private citizens or corporations are still allowed to monetarily gain from public resources.

In 79 AD, the Lucrine Lake oyster park was destroyed by the volcanic eruption that buried Pompeii. Much like the oyster park, Ancient Rome was eventually buried and forgotten. The grip of plague and sickness would cover much of Europe during the Dark Ages, and while the park was briefly revived, it was destroyed again by the Monte Nuovo eruption in 1538. It would be hundreds of years before the art of oyster farming would be revived once more.

DON'T FORGET
THE PEARLS

PEARLS ARE KNOWN AS THE "QUEEN OF JEW-els" and have been used for adornment and as a symbol of wealth throughout human history. Archaeological evidence dates some of the earliest natural pearl harvests to the Late Stone Age in 6000–5000 BC. The *Epic of Gilgamesh*, written around 2150–1400 BC in Mesopotamia, describes a hero who dives into the depths of the sea with weights on his feet for the "flower of immortality," a well-known reference to pearls. By 100 AD, Pliny the Younger declared pearls from the Persian Gulf as the most prized goods in Roman society.

Naturally occurring pearls were so valuable in antiquity, that in the year 69 AD, Roman emperor Vitellius financed an entire military campaign by selling just one of his mother's pearl earrings. Elizabeth I of England owned three thousand pearl-encrusted robes and eighty pearl-studded wigs.

These gems of the sea were found on every continent on the globe and are formed by the secretion of nacre, a shell-producing material from within the oyster's mantle tissue. Not all oysters produce shell using nacre. Most modern mollusks produce shell with calcium carbonate. It is estimated that at the historical peak, about one thousand oysters and mollusks were able to produce pearls using nacre. Today just a handful of these species still exist, due to overharvesting and ecosystem degradation.

Particular conditions are required for loose natural pearls to form, and this natural occurrence is very rare, leading natural pearls to be highly prized, valuable objects historically worn by kings, queens, rulers, and aristocrats.

Pearls were so rare that they were thought to have mystical properties. The Chinese thought they cured illness. Societies in India prized them as aphrodisiacs. During the Byzantine Empire, only the emperor was allowed to wear pearls. The ancient Greeks called them the "tears of the moon," and when the Gods cried, pearls were said to be their tears.

Natural pearls were typically collected by chance, stumbled upon from oysters gathered for food from their natural habitats. Demand for pearls drove a type of gold rush, ushering in the ancient profession of a "pearl diver."

Ancient pearl divers were willing to risk their lives for the prized oyster gems. The best divers could dive up to twenty-five meters in depth on a single breath. They would collect the most valuable oyster varieties, sometimes diving as many as forty times in a single day. Approximately one in a thousand oysters they collected would contain the prized naturally occurring pearl.

Historical sources of pearls collected in this way included the regions of the Gulf of Mannar (between India and Sri Lanka), the

Bay of Bengal, the Red Sea Egyptian coast, and the Persian Gulf along the Saudi Arabian coast.

Around the year 1284, thirty-year-old famed merchant explorer Marco Polo made an expedition into the area known as Malabar, India—"Maabar" in his travelogue—and documented the practice of pearl diving in his journal as follows:

You reach the great province of Maabar, which is not an island, but a part of the continent of the greater India, as it is termed, being the noblest and richest country in the world. It is governed by four kings, of whom the principal is named Sender-bandi. Within his dominions is a fishery for pearls, in the gulf of a bay that lies between Maabar and the island of Zeilan, where the water is not more than from ten to twelve fathoms in depth, and in some places not more than two fathoms. The business of the fishery is conducted in the following manner.

A number of merchants form themselves into separate companies, and employ many vessels and boats of different sizes, well provided with anchors. They engage and carry with them persons who are skilled in the art of diving for oysters in which the pearls are enclosed. These they bring up in bags made of netting that are fastened about their bodies, and then repeat the operation, rising to the surface when they can no longer hold their breath. And after a short interval they dive again. The greater portion of the pearls obtained from the fisheries in this gulf are round, and of a good lustre. The spot where the oysters are taken in the greatest number is called Betala, on the shores of the mainland.

The gulf being infested with a kind of large fish, which often prove destructive to the divers, the merchants take the precaution of being accompanied by certain enchanters belonging to a class of Bramins, who, by means of their mystical art, have the power of stupefying these fish, so as to prevent them from doing mischief. As the fishing takes place in the daytime only, they discontinue the effect of the charm in the evening; in order that dishonest persons who might be inclined to take the opportunity of diving at night and stealing the oysters, may be deterred. The enchanters are likewise skilled in the art of fascinating all kinds of beasts and birds.

*The fishery commences in the month of April, and lasts till the middle of May. The privilege of engaging in it is farmed of the king, to whom a tenth part only of the produce is allowed. To the magicians they allow a twentieth part, and consequently they reserve to themselves a considerable profit. By the time the period above-mentioned is completed, the stock of oysters is exhausted; and the vessels are then taken to another place, distant full three hundred miles from this gulf, where they establish themselves in the month of September, and remain till the middle of October. Independently of the tenth of the pearls to which the king is entitled, he requires to have the choice of all such as are large and well-shaped; and as he pays liberally for them, the merchants are not disinclined to carry them to him for that purpose.**

In North America, Native Americans harvested freshwater pearls by pearl diving into lakes and rivers in Ohio, Tennessee, and Mississippi. In the Caribbean, marine pearls were harvested by div-

---

* Marco Polo, *The Travels of Marco Polo*, ed. Manuel Komroff (New York: Boni & Liveright, 1926).

ing along the coasts of Central and South America, and Christopher Columbus saw West Indians diving for pearls on one of his voyages.

In 1579, King Philip II of Spain had a pearl of 250 carats, which originated from the island of Margarita off the coast of Venezuela. None that big have been found since. The largest pearl known from Pacific waters was seventy-five carats, found in 1882 in the Gulf of California.

In Australia, pearl diving began in the 1850s and remained strong until World War I and the invention of plastics, which replaced items typically made of shell.

In all of these locations, the occupation of pearl diving was typically reserved for enslaved people or captives. Many of the divers drowned, lost their lives to sharks, or lost hearing in both ears due to "swimmer's ear."

Officially known as external auditory exostoses, or EAE, "swimmer's ear" is a condition where bony masses grow inside the ear canal to protect inner ear components in individuals who spend prolonged time in or under the water. EAE has been found in fifty-thousand-year-old Neanderthal skulls and even in hominin populations dating back to 430,000 years ago, suggesting harvest and gather activities that required these individuals to spend enormous amounts of time underwater.

By the early 1900s, the natural pearl fisheries were exhausted and many countries involved in pearl diving lost significance in the market as cultured pearls became the dominant source of the oyster gems.

People in the Chinese Song dynasty (960–1279 AD) understood the process of making blister pearls on the inner shell surfaces of freshwater mussels, and this dynasty marks the advent of pearl farming. During the Ming dynasty (late thirteenth century AD), the blister pearl process was used to produce pearl Buddhas that were sold in temple markets.

The modern round pearl cultivation method known globally as the Mise-Nishikawa method was developed by "the father of modern cultured pearl production" Kokichi Mikimoto in Japan in the early 1900s. The Japanese cultured their first spherical marine pearls in 1907 using the Akoya pearl oyster species. Akoya pearls have been mass-produced since 1945.

Cultured pearl production now supports industries in more than thirty countries around the world, with China producing 98 percent of global cultured pearl output, of which freshwater pearls accounted for 99.5 percent, coming from the Unionidae freshwater mussel. The largest and most valuable of cultured pearls come from the silver- or gold-lipped pearl oyster *Pinctada maxima*, dubbed the white "South Sea pearl." Indonesia and Australia are its major producers. Japan has been the world's largest marine pearl producer for over a century and has developed cutting-edge technology in pearl oyster culture and pearl production.

Cultured oyster pearl production typically consists of five stages: oyster selection, implanting, nurturing, harvesting, and pearl processing.

All species of shellfish are capable of producing pearls; however, those of value and of interest as gemstones are limited to just two groups that secrete nacre (or "mother of pearl")—the marine pearl oysters of the family Pteriidae and the freshwater pearl mussels of the families Unionidae and Margaritiferidae.

After shellfish species selection, a nucleus (spherical polished shell-bead) is implanted into the specimen. Marine oysters are typically cultured for another one to two years before harvest, while freshwater mussels typically culture for one to five years before the pearl is harvested.

Upon harvest, the oyster or mussel is pulled from the water and the pearl is extracted from inside the shellfish. If the pearl is of quality design, the specimen can be reimplanted with a new nucleus and placed back onto the farm to culture additional pearls. If the pearl is of poor value, the specimen can be eaten and the shells can be used for mother-of-pearl beads or inlays.

HOW THE FRENCH
TAMED THE SEA

O N EVERY CONTINENT THAT OYSTERS HAVE
existed, humans overharvested the population to
extinction, beginning in ancient Egypt and spreading
with the ancient Greeks, Romans, and Asian cultures. As populations migrated northward into Britain, Ireland, and the Netherlands, new oyster beds were discovered and quickly devoured.

A charter in 1189 gave the town of Colchester, England, control over their prized wild "Colchester Natives" and kicked off the
oldest running oyster festival in the world, the Oyster Feast at Colchester, held annually since at least 1256.

As with most English residents, the notorious William Shakespeare was an avid oyster consumer, which eventually led him to
pen one of the most popular and repeated metaphors in human
history—that "the world is your oyster." Some would describe the
metaphor as an opportunistic reference, that the person is in a

position to take the opportunities that life throws their way. Others consider the metaphor to mean that your fate is in your hands, and if you're lucky, you'll find a pearl. But how does either of these suggestions truly relate to an oyster? Oysters simply stay put, and filter that which comes along.

The origin of the phrase dates back to the famous Shakespeare play *The Merry Wives of Windsor*, first published in 1602. In act 2, scene 2, characters Falstaff and Pistol argue over money as they enter a room;

FLASTAFF   *I will not lend thee a penny.*

PISTOL   *Why then, the world's mine oyster, which I with sword will open.*

FALSTAFF   *Not a penny.*[*]

Unlike the modern rendition we know today, Shakespeare's metaphor initially invoked a sense of violence as he suggests that if Pistol doesn't get what he wants, he will take it, by force if necessary. It takes force, persistence, and a bit of skill to open an oyster, and once mastered, we get what we desire. And if extremely lucky, sometimes inside we find more than we thought: a pearl, which has the power to be a life-changing event. It's a beautiful metaphor that has evolved over the last four hundred years to become one of the most popular and true metaphors for life.

During the Hundred Years' War of the 1330s, Europeans continued to consume oysters at unsustainable rates. It was not until

* William Shakespeare, *The Merry Wives of Windsor* (Washington, DC: Folger Shakespeare Library, n.d.), 2.2.1–4.

the extreme decline of the wild fisheries that France began to look to cultivation as a solution.

Unlike other countries (who created public commons for their seafood), France had deemed the wild oyster beds a national asset and sought to administer the bounty of the shorelines as the royal crown saw fit. In 1544, and again in 1584, French kings asserted the royal prerogative to own and farm the oyster beds in national interest, and even went so far as to unravel private fishing attempts on the oyster beds. Essentially, oysters became the property of the king.

King Henry IV (1553–1610) was said to eat three hundred oysters at a time. His grandson Louis XIV had them delivered fresh daily to Versailles (or wherever he happened to be) and was known for eating six dozen at a time. Oysters were considered the crown jewel at any royal event. Their importance was such that in April 1671 they cost a young chef his life.

François Vatel, a chef who oversaw extravagant events for Louis XIV, organized a two-thousand-person banquet in honor of the king at the Château de Chantilly. When he discovered his order of oysters and seafood was to be late to the party, Vatel took it upon himself to run himself through with a sword and die by suicide. Better to be dead than tell the party their oysters would be late. Vatel's body was found when the order of seafood finally arrived.

The French Revolution of 1789 saw the end of the French monarchy, but the oyster beds remained a national asset, at least in principle. Without a crown to oversee the reefs, poachers quickly seized upon the opportunity to harvest as many oysters as possible. By 1840, the French Navy was assigned to protect the remaining oyster beds around Arcachon; however, it was already too late. Places that had as many as fifteen oyster banks and prosperous fisheries, such as Saint-Brieuc, La Rochelle, Marennes, and

Rochefort, had been harvested to the point of no return, proving that it was indeed possible to overharvest the oyster beds beyond their ability to repair.

The French government dove into action and sent the brilliant embryologist Jean Jacques Marie Cyprien Victor Coste on a voyage along the French and Italian coast to inquire into the conditions of the fisheries. It was his mission to see how the oyster fisheries could be artificially aided.

Jean Victor Coste had become the professor of embryology at the Collège de France in 1841. He was a renowned scholar and medical doctor and was close to the ruling elite of the Second Empire. Charged by Napoleon III with his fact-finding mission, Coste immediately set out for the famous Sergius Orata farm site in the Naples area.

Almost one hundred years earlier in 1764, King Ferdinand IV of Naples had revived the oyster farm of Sergius Orata in Lake Fusaro. Coste visited this farm on his journey along the coast and made detailed investigations into the techniques. The Lake Fusaro system of cultivation was strongly recommended by Coste upon his return to France, and experiments were at once started to prove whether the techniques were applicable to French waters.

In his 1858 report to the French king, Coste writes:

To this deplorable state of matters there is one remedy, of easy application, of certain success, and which will give an incalculable supply to public nourishment: this remedy consists in undertaking, at the expense of the State, under the care of the Marine Administration, and by means of its vessels, the sowing (with oysters) of the shores of France, so as to restock its ruined beds, to revive those which are extinguished, to extend those which prosper, to create new ones wherever the nature

*of the bottom will permit their establishment; and when,
through this generous commencement, the productive beds
shall have sufficiently developed themselves in all places, they
might then be submitted to a salutary system of regulated
gatherings, allowing some to remain quiet while others are
worked—an arrangement which for a century has preserved
the beds of Cancale and Granville from destruction.*

These actions constitute the first time a "government" agency
in history had stepped in to protect, conserve, and restore a diminished natural resource. An action that seems all too common today.

Most importantly, Coste scientifically proved that the breeding techniques of Orata were the most efficient method of oyster
breeding. Artificial insemination would be impossible due to their
complex fertilization and incubation method, therefore oysters
had to be naturally bred with each other. Natural breeding methods are still carried out today.

Coste asked Napoleon III for 8,000 francs to restock the Bay
of Saint-Brieuc. He imported oysters into the bay, hired a boat to
guard them, and replicated the works of Sergius Orata. Despite the
deep waters and strong winds, Coste immediately saw success with
his results.

The bundles of branches and structures Coste built soon
became covered with oyster spat. Within six months of the planting, the success of the Bay of Saint-Brieuc experiment was assured.
An estimated twenty thousand baby oysters settled on each
collector. A year later in 1859, Coste recommended to Napoleon
that the entire French coastline, and even the colonies of Corsica

---

* Thomas Campbell Eyton, *A History of the Oyster and the Oyster Fisheries* (London: John Van Voorst, 1858), 37–38.

and Algeria, follow his lead. Napoleon instituted the request, and the flat Belon oyster flourished in the region once again.

Coste went on to form a partnership with the captain of the port of Concarneau and created the Concarneau "fishpool-laboratory" in 1859, which holds the title of the oldest marine research institute in the world. Coste eventually became general inspector of river and sea fishing for France in 1862.

To this day, most farms in France continue to use the Sergius Orata–inspired bottom-planted method. Most oysters are spawned and grown in the traditional way, planted directly on the bottom sediment of a small leased tidal pool. Modern oyster farms continue to use leased tidal pools but have begun to use bags or containers placed on the bottom sediment to contain the oysters. The containers help reduce predation and make handling the oysters more efficient.

The success of Coste and other Frenchmen undoubtedly led to the revival of the Belon oyster in the French region; however, it was a chance event that led to the more robust Portuguese (*Crassostrea angulata*) oyster populating the French coastline.

Though the Portuguese oyster was originally considered a native oyster to Europe, genetic testing has now revealed it originated from the Pacific coast of Asia. Imports of the Portuguese oyster to the French Arcachon Basin began in the 1860s in response to the dwindled Belon oyster populations. One of these merchant vessels crowded with Portuguese oysters found itself caught in a violent storm. In order to escape inevitable death, the vessel was forced to dump its cargo into the Gironde estuary near the town of Bordeaux, France.

The escaped Portuguese oysters flourished in the region and soon represented two-thirds of all oysters farmed in France by the 1900s. By the 1920s, the much more fragile Belon oyster was nearly wiped out by disease, and by the 1960s, just 20 percent of French production was the Belon oyster, while Portuguese oysters represented 80 percent.

This all changed in the 1970s, when another wave of disease swept the French coastline, this time killing all Portuguese oysters. Following the carnage, the Japanese Pacific cupped oyster (*Crassostrea gigas*, a.k.a. *Magallana gigas*) was introduced to the region. Today, this is the most widely cultivated species in France and in the world.

John Singer Sargent, *The Oyster Gathers [sic] of Cancale*

JAPAN, TO
THE RESCUE!

CULTURED FOR HUNDREDS OF YEARS, THE JAP-anese oyster (*C. gigas*, referred to as the Pacific oyster) is easy to grow, tolerant to various environments, and easily spread from one region to another. The global domination of the Japanese Pacific oyster began in 1920, enabled by the rise of oceanic trade ships and two young Japanese immigrants working on the West Coast of the United States.

Around 1900, Japanese immigrants supplied most of the labor to the American West Coast oyster industry, not only because of their low wages but also due to the experience they brought with them from the much more developed Japanese oyster industry.

Two young Japanese men, J. Emy Tsukimato and Joe Miyagi, residents of Olympia, Washington, educated in and graduated from the public schools in Olympia, had both worked with oysters on the beds of Oyster Bay near Olympia during summer vacations.

It was there that they dreamed of transplanting Japanese oysters from their native country into the Puget Sound.

The men conducted research and eventually settled on Samish Bay, near Blanchard, Washington, as the site of their new oyster farm venture. They formed a company, raised significant capital, and purchased six hundred acres of farmland from the Pearl Oyster Company, which at the time struggled financially trying to farm native Olympia oysters.

Experiments to ship Japanese Pacific oysters to the West Coast of America started as early as 1902, but mortality rates were very high. In April 1919, Tsukimato and Miyagi contacted exporters from the Miyagi region of Japan and imported four hundred cases of Pacific oysters. There is speculation that the two men must have had economic or family ties to the Japanese oyster industry, for this was considered a major export to America at a time when access to Japanese oyster seed was reserved for Japanese farms only.

The Japanese seed traveled sixteen days aboard a cargo vessel, constantly kept moist by the ship's crew as directed by Tsukimato and Miyagi. Like immigrants arriving on the shores of the United States, the oyster seed was subjected to inspection by the U.S. customs officer of Seattle, and then was taken by a scow vessel to the oyster acres in Samish Bay to be planted.

All of the adult oysters died; however, the spat (young oysters) attached to the outside of the adult oyster shells survived and grew! This observation led the two men to conclude that this was the best method to ship oysters across the sea, as young spat attached to shell. Additional shipments were successful based off of these observations. Tsukimato and Miyagi took great care in reducing mortality, and the oysters grew rapidly. Soon they had a flourishing business and had successfully established the Japanese oyster on the West Coast.

Unfortunately, their success was quickly stripped. The legislature of the state of Washington during its session in 1921 passed what is known as the Alien Land Law. This law prevented the ownership, or leasing, of land in the United States by certain immigrants, which prevented the ownership of the Samish Bay oyster farm by the Japanese men.

In 1923, Tsukimato and Miyagi were forced to sell their farm operations, which were purchased by their Anglo-Saxon friends and oyster business associates E. N. Steele and his brother Charles Steele. The four men maintained a cooperative venture for years to come and worked diligently to convert West Coast oyster lovers to the newly established Japanese oysters, marketed as "Rock Point" oysters.

The team built a special truck for hauling and showcasing the oysters across the West Coast and as far east as Salt Lake City, offering tastings at restaurants and grocery stores. Slowly but surely, they created a loyal following.

E. N. Steele became a legend in the West Coast oyster industry. In August of 1930, he helped to create an organization to grapple with issues of price wars and size standards, calling it the North Pacific Oyster Growers Association and dubbing the Japanese oyster the "Pacific" oyster. In 1934, the organization changed names again, to the modern-day Pacific Coast Shellfish Growers Association.

The American West Coast and Japanese oyster industries suffered greatly during the outbreak of World War II. International shipment of oyster seed from Japan was halted by American warships. West Coast American farms dependent on Japanese seed imports scrambled to find a new source of seed.

Dick Steele, the son of E. N. Steele, returned home from the war in 1945 and went to work for the family oyster business. His assignment was to find growing areas along the West Coast with natural sets of Pacific oysters. His search led him to Dabob Bay on Washington's Hood Canal, where his father had purchased a lease years earlier. In 1946, the natural Pacific oyster sets in Dabob Bay met expectations, and they began to sell the seed oysters to growers around the region. The West Coast industry began to recover, yet more sources of seed were still greatly desired. As soon as the war ended, it was oyster seed that was first exported from Japan.

In the 1950s, the Japanese Pacific oyster was unintentionally introduced to New Zealand, most likely in the ballast water and hulls of ships. Farmers began to notice the Pacific oyster outcompeting the native Sydney rock oysters, growing faster and producing thicker meats. Since a well-established market already existed throughout Asia, farmers switched to the Pacific oyster, which is now the dominant species grown on Australian and New Zealand oyster farms.

In 1966, the Japanese Pacific oyster was imported to France to replace the decimated Portuguese oyster population. Major studies were undertaken to understand the environmental ramifications of replacing an oyster species. Faced with a quickly declining wild stock (due to disease) and a collapsing oyster industry, the French government gave the go-ahead. Within a decade, French production of the Pacific oyster had shattered historical records as the faster-growing, more-productive species thrived along the French coast.

In addition to the importation of the Japanese oyster, the French farmers also imported Japanese husbandry techniques, including the rack-and-bag method of farming (in which oysters grow in bags attached to racks planted on the bottom sediment). This new style of culture allowed the French industry to expand into growing areas prohibited by traditional bottom-planting culture, providing a boom in production.

Due to the prolific nature of the Pacific oyster, it has been used to replace the dwindling native oyster populations around the world and start oyster fisheries where none had existed before. Pacific oysters have even continued to invade areas by hiding in ship ballasts and hulls. In the year 2000, Japanese Pacific oysters accounted for 98 percent of the world's cultured production. In 2003, global Pacific oyster production was worth $3.69 billion!

Much good has been created by the worldwide proliferation of the Japanese oyster; however, it is important to mention the unintended consequences. The ocean no longer contains isolated environments, but is instead one big mixture of creatures.

International shipping trade and the globalization of the Japanese oyster has spread a host of "invasive" animals around the globe. Three hardy oyster predators, once native to Japan, now occupy every coastline of the world: the Japanese oyster drill, flatworm, and parasitic copepod. Each of these predator species was able to hitch a ride via exported Japanese oysters and each species now calls every corner of the globe its home.

As farmers we must contend with these predators. Even farms on the East Coast of America (where Pacific oysters are prohibited) have been invaded by these predators' supreme survival traits. Their impact on the global oyster industry is very real, and their presence has dictated to much extent what form of farm culture is used on a particular site.

THE WEST COAST

T HE WEST COAST OF AMERICA IS HOME TO THE one and only *Ostrea lurida* (Olympia oyster), found clinging to the rocks along the coastlines from southern Alaska to Baja California in Mexico.

The colder waters of the Pacific Ocean deter the Olympia oyster from spawning in much of the region, constricting the bivalve to survive naturally in only the warmer inlets and the more southern waters of the West Coast. Much of the Northwest contains a mud sediment environment, which further hinders the success for oysters. These impediments restricted the real estate available for oysters, while providing a flourishing habitat for a variety of clam and mussel species.

From Alaska to California, shell middens (giant piles of shells discarded after consumption) left behind by Native Americans show the bulk of species eaten to be mussels (*Mytilus californianus*, *Mytilus trossulus*), butter clams (*Saxidomus giganteus*), littleneck clams (*Protothaca staminea*), cockles (*Clinocardium nuttallii*), whelks (Nucella species), and barnacles (Balanus species), demonstrating a reliance on the various species of shellfish as a staple food source for Northwest Coast First Nations for at least five thousand years. The minimal consumption of oysters in the Northwest Native American diet is due to their natural scarcity in the environment.

Recent findings suggest the Native populations did more than just harvest clams for sustenance; they created clam gardens. Constructed in the late Holocene Epoch, clam gardens consist of human-engineered intertidal terraces, made by building rock walls in the low intertidal soft sediment to stabilize sediments at a specific height to enhance shellfish productivity. Within these "gardens" clams were encouraged to propagate, sediment was augmented to benefit the species, and selective harvests were conducted to maintain population stability.

In this way, Northwest Coast First Nations were able to sustainably manage their shellfish populations well into the late 1800s, about the time when Europeans began the infamous West Coast pilgrimage.

On the western coast of Central America, shellfish harvests by Native Americans are well documented by shell middens riddled along the coastline, the majority of which are comprised of mussels and barnacles. As the great Inca Empire came to absorb these populations, they too acquired the skill and taste for shellfish.

South American tribal populations also relied heavily on shellfish for sustenance. The Chono, Alacaluf, and Yamana Indians occupied the entire Chilean coastline southward into Cape Horn. These southernmost tribes were known as the shellfish gatherers. As with their northern counterparts, the rugged terrain of boulders, fjords, islands, and mountains made travel inland extremely difficult and hazardous. Dense forests made horticulture impractical, and the land was sparce of large game and edible plants. Travel was done mainly by canoe, and food sources consisted of mostly seaweeds, shellfish, beached whales, and seals.

It's important to note that the human occupation of the west coast of the Americas dates back much further than that of the east coast. Most migration academics conclude that human expansion from the land bridge connecting modern day Russian with Alaska allowed the first wave of humans into the Americas sometime between 26,000 and 19,000 years ago. Following the "kelp highway" and reliable sources of shellfish, humans most likely used ancient canoe-type craft to transverse the coastline, and by 14,500 years ago, human occupation is found to have extended as far south as Monte Verde in southern Chile.

In comparison, the oldest-known South American east coast human skeleton is estimated to be 10,400 years old, found inland from the coast in a river shell midden in Capelinha, Brazil. Radiocarbon dating of caribou collagen is evidence of human arrival in the New England region by about 12,530 years ago, with the oldest confirmed sites in Maine and the Connecticut River Valley.

The lack of large naturally occurring oyster beds on the West Coast created immense opportunities for distribution of the Eastern oyster. Since the early 1800s, Americans have tried to cultivate the native East Coast oyster on the West Coast without much success. It was found that the journey by steam train took too long, and if the oystermen were actually lucky enough to plant the Eastern oysters into the West Coast waters, the majority of oysters would die due to extreme variations in environment.

Any Eastern oysters that did survive in the Pacific Ocean farms found the waters much too cold to spawn. It was deemed that a constant supply of East Coast oyster seed would be needed to maintain a presence of *Crassostrea virginica* on the West Coast, and soon all ventures were abandoned.

Nonetheless, Americans striking it rich in California longed for the oysters of their East Coast memories and were willing to pay top dollar for a taste. Attention turned to the West Coast native Olympia oyster populations and their numbers were quickly decimated as gold-striking Americans consumed all they could.

San Francisco Bay's entire Olympia oyster population was harvested in a matter of years. Willapa Bay in Washington soon became the heart of the West Coast oyster industry. Oysters were packed in barrels of seawater and shipped to the gold mining fields where California's first famous cuisine, the Hangtown fry, became a mark of prosperity for gold-rich miners and the status symbol of the day, costing as much as six dollars for a meal (a fortune in those days). A few drinks and a Hangtown fry were considered a gentleman's evening. What was a Hangtown fry? Fried breaded oysters, eggs, and fried bacon, cooked together like an omelet.

THE EAST COAST

CULTIVATION OF *CRASSOSTREA VIRGINICA* (referred to as the Eastern oyster) traces back to the Native American population. Shell middens sprinkled the riverbanks and East Coast shorelines wherever the colonists went and were constantly described as landmarks for the early explorers.

Archeological research reveals that East Coast Native Americans feasted heavily on oysters for nutrition, yet they managed to find balance between the amount they harvested and the amount they allowed to remain in the wild. They did not farm shellfish, but instead they sustainably managed their wild fisheries.

Middens dating to the Late Archaic Period (8000–1000 BC) are disproportionately made up of oysters over any other species of shellfish. Starting around 3000 BC, evidence of large-scale exploitation of oysters is present along modern-day coastline settlements. There is evidence to suggest any coastal Native American sites older than 3000 BC were buried under the ocean as sea levels rose.

Interestingly, we see a decline in shellfish harvests during the Woodland Period (1000 BC to 800 AD). Some theories suggest this is due to a lowering of ocean levels, causing the coasts in some areas to extend much farther away from current locations. During this period, many of the historic sites are abandoned or see a large decline in population.

By the Mississippian Period (800–1600 AD) coastal shore-lines resumed their previous levels and many of the Late Archaic Period sites experience a renewal in occupation. Exponential Native American population growth is measured during the Mississippian Period, most likely due to in-migration from other regions, the establishment of a regional settlement system, and the eventual adoption of maize agriculture. Enormous harvests of shellfish, mostly oysters, in numbers much larger than those during the previous Late Archaic and Woodland periods, present evidence of flourishing Native American populations, which reached record numbers just prior to European arrival.

The Native American population crashed as the first-contact European explorers introduced old-world disease to the Americas. By the time of mass immigration to the continent, the Native American population had been reduced to just a fraction of their peak population numbers. This abrupt reduction in harvest pressure is evidenced through shell middens, and allowed the surviving oyster reefs to quickly rebound and thrive in just a decade or two, leaving almost pristine reefs by the time of mass colonist immigration.

When left to their own devices, oysters will form giant reef systems packed with trillions of individual oysters. These reefs have the ability to protect shorelines from tidal forces associated with rising oceans and superstorms. Like coral reefs, these complex reef systems also provide habitat for a wide range of ocean biodiversity and consume algae blooms, maintaining proper water clarity and quality for marine organisms to thrive.

Regardless of the location along the East Coast, the vast majority of oyster reefs in America remained intact by the time of mass European settlement. When colonists first came to America, the oyster reefs were so massive that they had to be drawn on maps to keep the wooden vessels from striking them and sinking. Francis Louis Michel, a Swiss visitor, wrote in 1701, "The abundance of oysters is incredible. There are whole banks of them so that the ships must avoid them. A sloop, which was to land us at Kingscreek, struck an oyster bed, where we had to wait about two hours for the tide. They surpass those in England by far in size, indeed they are four times as large. I often cut them in two, before I could put them into my mouth."[*]

Crassostrea virginica gets its name from its discovery by the original Roanoke, Virginia, colony of 1583. Early settlers recounted the oysters being about a foot in length.

In 1607, the first permanent English settlement in America was established as the colony at Jamestown on a peninsula in the Chesapeake Bay. By 1609 the dwindling colony was on the verge of starvation, and it was oysters that the colonists survived on during this time.

The infamous Captain John Smith was amongst those first colonists in Jamestown, and in 1608 he gathered fourteen men to join him on a voyage up the Chesapeake. The word *Chesapeake* derives from the Native American word *Tschiswapeki*, which translates into "the great shellfish bay." In John Smith's diary, he is quoted as saying that the oysters in the Chesapeake Bay "lay as thick as stones." It's

---

[*] James Wharton, *The Bounty of the Chesapeake* (Charlottesville: University Press of Virginia, 1957).

important to note that he could obviously see the oysters blanketing the bottom of the bay. Today, without this giant mass of oysters to consume the algae blooms, the bottom of the bay is not viewable and lies obscured by a cloudy haze of phytoplankton.

Describing his Chesapeake Bay oyster account in 1612, William Strachey wrote, "Oysters there be in whole banks and beds and those of the best. I have seen some thirteen inches long. [The Indians] . . . hang the oysters upon strings . . . and [dry them] in the smoke, thereby to preserve them all the year."*

In the 1620s, pilgrims in the Plymouth colony were known for feeding their hogs wild clams and mussels, while keeping the oysters for themselves. Roger Williams's 1643 treatise on the language of the Narragansett Indians noted that during the summer months the Native Americans would wade and dive deep for shellfish. By 1680, settlers in Maryland complained of hardships so severe that "their supply of provisions becoming exhausted, it was necessary for them, in order to keep from starvation, to eat the oysters taken from along the shores."†

One of the most long-standing Native American traditions associated with shellfish has been their wampum, a type of jewelry that consists of beautiful handmade white shell beads from the North Atlantic channeled-whelk shell, and white and purple beads made from the quahog or hard-shelled clam. Archeological excavations in New York have turned up beads from before 1510; however, the tradition is thought to date back much further.

Wampum beads were tubular in shape, about a quarter of an inch long and just an eighth of an inch wide, with a hole through

---

* Victor S. Kennedy and Linda L. Breisch, *Maryland's Oysters: Research and Management* (Maryland Sea Grant, 2001), 100.

† Kennedy and Breisch, 100.

the center for sewing. Native American women typically made the beads by rounding out small pieces of shellfish shell and then piercing a hole through the center and stringing them together.

Wooden pump drills with quartz drill bits and steatite weights were used to drill the tiny holes into the shells. The beads would be strung together with deer hide thongs or other fibers. The wampum beads were then woven into a belt like tapestry, as a way to convey important stories, ceremonies, or traditions. These belts could contain thousands of small wampum beads, and the finished products were coveted by Native Americans as one of their most prized and valuable possessions.

Only the coastal Native American tribes had access to the prized shellfish shells used for wampum production, which made the wampum scarce and valuable, especially to landlocked tribes. Early Europeans quickly recognized the value of wampum and began trading them as a type of currency. The trend caught on, and wampum became America's first produced currency. Wampum was legal tender in New England from 1637 to 1661 and briefly became legal tender in North Carolina in 1710, but died out as a common currency in New York by the early eighteenth century.

Europeans' metal tools revolutionized the production of wampum and by the mid-seventeenth century, workshops mass-produced tens of millions of beads into circulation. Eventually, the vast majority of wampum beads in circulation were created by the English and Dutch production companies, which ultimately reduced the value of the wampum to a status of obsolete.

The social and cultural significance of wampum remains to this day. Most tribal treaties or events were documented on wampum belts, some of which continue to be passed from generation to generation. Even George Washington had a wampum belt created to signify the unity of the tribal nations and the United States of America. Today, many of the coastal tribes of New England continue to craft wampum jewelry, much of which is sold to high-paying tourists looking to take a piece of New England Native American history home with them.

Like their clam and whelk shell counterparts, oyster shells also had higher value as a raw material, specifically for the manufacture of lime used in masonry mortar.

Limestone is not readily available in New England, so oyster shells, made of calcium carbonate (the same material as limestone), readily took the place of limestone in all masonry products like cement, fertilizer, paint, and road materials. Early colonial America was literally built out of oyster shells.

The oldest use of oyster shells in construction is called tabby concrete, an ancient form of structure building brought over with the Spanish in the sixteenth century. Tabby's origin is African, but little else is known about when or where it was invented. The technology is simple; oyster shells are crushed and burned to create quicklime, then mixed with water, sand, ash, and broken oyster shells to form a mortar-type mixture that can be used to build foundations, floors, walls, and roofs. Native American shell middens served as the primary source of oyster shells, which Spanish settlers quickly mined for production.

Juan Ponce de León led the first known expedition into Florida in 1513 and named the area La Florida during this first voyage. He is credited with making the first contact between Europeans and the Native American Calusa Indians when he landed somewhere along Florida's eastern coast, then traveled down to the Florida Keys and north along the Gulf Coast, perhaps as far as Apalachee Bay on Florida's western coast. In 1521, Ponce de León returned to Florida in an attempt to establish a Spanish colony, but the Calusa Indians resisted. Ponce de León was shot in his thigh by an arrow and eventually died from his wounds.

The Calusa Native Americans had lived in the area for more than two thousand years, and more than twenty thousand Calusa people lived in the area during the mid-1500s. The Calusa ruler, King Caalus, lived in a large house on Mound Key, an island created from disposed clam and oyster shells.

When Pedro Menéndez de Avilés was appointed Spain's first governor of Florida in 1566, he marched two hundred soldiers to Mound Key and in 1567 established the San Antón de Carlos fort, the oldest surviving example of tabby concrete construction in North America. Many of the Calusa people died from disease brought by the Spanish, and in 1569 the fort was abandoned due to another uprising by the Calusa people.

Tabby concrete building techniques spread quickly throughout early colonial America. Learning from the Spanish tabby methods, the British quickly adapted the construction technique to their Beaufort, South Carolina, sites and by 1700 the tabby concrete material was being used as far north as Staten Island, New York, as demonstrated by the still-standing Abraham Manee House, erected circa 1670.

Brunswick Town was founded in 1725 as the first permanent European settlement on the Lower Cape Fear River of southern North Carolina and utilized oyster shell mortar as the foundations for their homes. Even the first Cape Hatteras lighthouse, which operated from 1802 to 1871, used an oyster-based mortar.

So many New Yorkers burned oyster shells in their homes for construction that the city passed a law in 1703 outlawing the burning of oyster shells due to public health concerns. House fires and harmful exhaust from the burning of shells were common.

By the early 1700s, harvest of oysters exclusively used for raw material lime production was so wasteful that the Rhode Island colonial assembly outlawed the practice by statute in 1734, noting the unacceptable waste of oyster meats (for food) as unshucked oysters were fed into the production kilns.

As the Dutch began to migrate into the New World, early Dutch settlers were disappointed that the oysters found along the East Coast were not of the pearl-producing kind; however, they took advantage of the abundance of oysters across the "New Amsterdam" region.

New York City (New Amsterdam) was a mecca for wild oysters. The first colonists wrote stories of massive shell middens that lined the riverbanks, a hallmark of the gluttonous Native American oyster feasts. Ellis and Liberty islands were originally called "Little Oyster Island" and "Great Oyster Island" because of the beds surrounding them. And New York's Pearl Street was originally named after a shell midden found nearby and was later paved with oyster shells.

The convergence of the Hudson and East rivers produced some of the most productive oyster grounds in the world. In the early 1800s, more than half the world's supply of oysters was contained in just 350 square miles of New York's harbor alone. Millions were exported annually to Europe in the early colonial period.

The original New York oyster population was capable of filtering all of the water in New York Harbor in a matter of days. However, it was soon harvested into near extinction. In 1658, the New Amsterdam Dutch Council had already started limiting the hours and locations oysters could be gathered due to overharvesting. As early as 1704, residents of Rockaway attempted to regulate the taking of oysters in their waters to locals only.

Oyster meats were the original bar snack and a New Yorker's preferred type of protein. Unlike the plates of shucked oysters on the half shell we've grown accustomed to today, oysters in the late 1700s would have been served as a broken-off clump of the reef all fused together, with a hammer and knife for the patron to use to open the oysters for themselves. Piles of discarded oyster shells were the original skyscrapers of the city. New York City should have been called "The Big Oyster."

For a more detailed history lesson on New York and the oyster, I highly recommend Mark Kurlansky's book, *The Big Oyster: History on the Half Shell*. Kurlansky's research is thorough and highly entertaining.

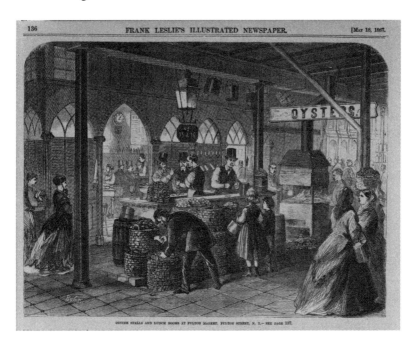

Albert Berghaus, *Oyster Stalls and Lunch Rooms at Fulton Market, Fulton Street, N.Y.*

OYSTER STANDS IN FULTON MARKET.—[Drawn by A. R. Waud.]

Alfred R. Waud, *Oyster Stands in Fulton Market*

AMERICA'S APPETITE FOR
THE OYSTER GROWS

AS DEMAND FOR OYSTERS INCREASED, THE efficiency of harvesting improved. Native Americans harvested oysters by hand, wading or diving into the water and using minimal tools to break the oysters off the reef. This simple form of harvest minimized damage to the oyster reef and the surrounding ecosystem, which helped sustain the oyster populations for centuries.

By the 1700s, the Europeans had introduced the harvest method of tonging for oysters, in which oystermen in boats would drift across the surface of the water and use long wooden or metal rakes to reach down and break off chunks of wild oysters from the reef below. This method increased productivity but also damaged great portions of the reef. Instead of targeting specific oysters for harvest as the Indians had done, the method of tonging was more of a take-all type of approach.

Reginald Hotchkiss, *Oyster Tongers, Rock Point, Maryland*

Oysters grown in deepwater reefs were the most prized for their cold, clean environment and high-salinity taste that made them unlike any oyster the Europeans had been accustomed to. The method of tonging allowed the deepwater oyster reefs to be more easily harvested, which quickly lead to their extinguishment.

An account from Ernest Ingersoll's report, *The Oyster Industry*, published in 1881, gives us insight into the industry during the American Revolution:

> *The coming on of the war of secession found the Boston oyster-trade in its most flourishing condition. More cargo-oysters were brought then, than ever since; prices were high and profits large. The shipping interests fostered by it were large, too, for the competition of railways and steamers had hardly made itself felt. Most of the large dealers ran lines of vessels of their own, as well as chartering additional assistance in the spring.*

*In the demand for fast sailers which the oyster-business created,
is found the origin of that celebrated model of sailing vessel that
made America famous on the seas—the clipper-ship. The first
of these were made by Samuel Hall, a noted ship-builder, at his
yard in East Boston, and were named Despatch, Montezuma,
Telegraph, and Express. They were from 90 to 120 tons, old
measurement, and carried an average cargo of 2,500 bushels
of oysters. Six months in the year these clippers were devoted to
bringing oysters from Virginia. There were thirty-five or forty
of these "sail" running, and in the summer they would go fishing. The freight tariff on oysters was then 20 cents, and during
the war it went as high as 25 cents a bushel. The war interfered
sadly with the business of oystering. Often the military operations did not admit of the cultivating and raking of the beds
in Virginia and Maryland, or of the schooners from northern
ports going where they wished to buy.**

By the 1810s, dredging began to replace tonging as the harvest
method of choice. Dredging was much more efficient than tonging
as it utilized a metal net that was dragged along the bottom of the
seafloor by a boat. This only accelerated the overharvest and the
depletion of the deepwater oyster beds in New England.

By the early 1800s, most states had passed legislation to protect
the remaining shallow-water oyster beds by making harvests available only to state residents. The wild oyster fisheries continued to
ramp up, yet in a counterintuitive move, Rhode Island issued the
first private harvest grant in June of 1798 to Samuel Thurber, who
petitioned the General Assembly for a two-acre grant near Sabin

---

* Ernest Ingersoll, *The Oyster Industry* (Washington: Government Print Office,
1881).

Point for the purpose of cultivating oysters.

Presumably this was the first private oyster farm in America, and Thurber was issued a renewable charter for the term of six years that forbade the general public from "molesting or disturbing the said Samuel in his enjoyment of the provisions of his charter."* Upon the expiration of his charter in 1804, Thurber did not renew his oyster lease. Apparently Thurber found farming oysters much harder than harvesting wild ones.

With oyster demand increasing and the New England oyster beds exhausted, New England oystermen migrated to Maryland and Virginia to harvest oysters from the great Chesapeake Bay, one of the most productive oyster grounds in the world.

Virginia passed "an Act, to prevent the destruction of Oysters within this Commonwealth"† in 1811 in response to the influx of out-of-state oystermen. The law prohibited using "any drag, scoop or rake, or other instrument, except tongs" to harvest oysters. Unintended consequences of the legislation spurred regional conflict by sending more dredgers into Maryland. In 1820, Maryland passed "An Act to prevent the destruction of Oysters in this state."‡ It banned residents from others states from harvesting oysters in Maryland waters and also outlawed dredging.

---

* Michael A. Rice, "A Brief History of Oyster Aquaculture in Rhode Island," in *Aquaculture in Rhode Island: 2006 Yearly Status Report* (Coastal Resources Management Council, 2006), 24.

† *Supplement, Containing the Acts Passed at a General Assembly of the Commonwealth of Virginia, of a Public and Generally Interesting Nature, Passed Since the Session of Assembly Which Commenced in the Year One Thousand Eight Hundred and Seven* (Richmond: Samuel Pleasants, 1812), 76.

‡ "Department of Natural Resources," Maryland Manual On-Line, msa.maryland.gov/msa/mdmanual/21dnr/html/dnrf.html.

Bushels, the standard one hundred count of oysters, were being harvested around the clock by lawbreakers. Deemed "oyster pirates," these out-of-state oystermen cared little for the new laws and outnumbered the authorities so greatly that penalties for their crimes were almost nonexistent. In 1839, seven hundred thousand bushels of oysters were harvested in Maryland, and by 1850 that number had doubled to 1.4 million bushels.

Oyster Pile, Hampton, Va.

Oyster mania spread like wildfire to the West Coast on the heels of the Oregon Trail, the California Trail, and the Gold Rush. The advent of canning technology helped increase consumption of oysters as they could be preserved long enough to be eaten by far-away markets. By the 1850s, Baltimore canned at least four million bushels of oysters annually, exported via railroad as the rail network expanded across the country. West Coast gold miners were finally able to consume the East Coast oysters they craved.

Following the surge in oyster fisheries in the Chesapeake, Rhode Island passed the state's first aquaculture law, the Oyster Act of 1844, which established a system of leasing tracts of submerged land to private individuals for the purpose of culturing oysters as well as a board of three shellfishery commissioners.

One of the first leases permitted under the Oyster Act was to Robert Pettis of Providence, who by 1890 became one of the largest leaseholders in the state. Privatization of the ocean was contentious. Fishermen became openly rebellious, stealing oysters and vandalizing private lease beds. Multiple arrests were made and a number of court cases in the 1850s upheld the power of the General Assembly to grant private leases.

By the late 1860s, a Chesapeake Bay oyster dredging captain made $2,000 a year, when most Marylanders earned just $500 or less. The end of the American Civil War in 1865 created a lot of disposable income for many Americans, and they began to look for status foods. Oysters were at the top of this list, driving demand to historically peak numbers.

Illegal harvesting was rampant as oystermen clashed across multiple fronts: the locals versus the out-of-staters; dredgers versus tongers; Marylanders versus Virginians; law-abiding citizens versus the oyster pirates. An all-out oyster harvest frenzy consumed the Northeast and Chesapeake Bay regions. People resorted to guns and violence. Mayhem ensued. Fortunes were made. Oystermen turned up dead.

THE

OYSTER WARS

MARYLAND CREATED THE STATE OYSTER Police Force on March 30, 1868, in an attempt to regain control of the waterfront. Commissioners of the force, which became known as the "Oyster Navy," were given high salaries, boats, and the authority to purchase guns and ammunition.

Captain Hunter Davidson led the Oyster Navy and was "required to keep his vessel constantly cruising, when circumstances will permit, wherever opposition to the oyster law has arisen, or is likely to arise, within the boundaries of the State, and that every locality where a violation of the law exists, or is likely to arise, shall be visited as often as the duties of the force and the conditions of the vessel will permit."*

---

* "State Oyster Police Force," Maryland Department of Natural Resources, March 30, 1868, dnr.maryland.gov/documents/OysterPoliceAct_1868.pdf.

Armed with a slew of weapons, including a howitzer and a 1,450-pound cannon capable of firing twelve cannonballs, the Oyster Navy chased down oyster pirates and sank their boats when necessary. Captain Davidson became a target and narrowly avoided an assassination attempt in 1871.

The Oyster Navy continued to operate into the early 1900s; however, it was never successful in controlling the illegal harvest of oyster beds in the state. Overharvests led to the eventual decline of the industry starting in the early 1900s, though the Oyster Wars continued well into the twentieth century. In 1959, the Potomac River Fisheries commissioner H. C. Byrd ordered the fisheries police to disarm after an officer killed a Virginia waterman who was dredging illegally. The move was credited with bringing an end to the violent conflicts.

While chaos consumed the wild oyster fishery farther south, the success of the privatized aquaculture industry in Rhode Island had blossomed into a multimillion-dollar operation with annual lease fees paid to the state exceeding $120,000. Leases for aquaculture peaked in 1911 at about twenty-one thousand acres, or about 20 percent of the entire bottom of Narragansett Bay.

The advancement of the Rhode Island aquaculture industry ushered in scientific research to aid methods of improved production. In 1888, scientists established the Rhode Island Agricultural Experiment Station, and in 1892 the Rhode Island College of Agriculture and Mechanical Arts (forerunner of the University of Rhode Island) was founded.

Oyster farmers in Point Judith Pond noticed production there was declining, while, at the same time, production in Narragansett Bay was increasing, and they brought these observations to college president Josiah H. Washburn, who in response authorized the establishment of Rhode Island's fist marine laboratory

Schell and Hogan, *The Oyster War in Chesapeake Bay*

in the village of Jerusalem on Point Judith Pond in July 1896. This is the third-oldest marine laboratory in the United States, only preceded by the U.S. Fisheries Commission Lab founded in Woods Hole, Massachusetts, in 1875, and the Marine Biological Lab, also founded in Woods Hole, by Harvard naturalist Louis Agassiz in 1888.

The Jerusalem lab has since been demolished, but the original site is now the Coastal Fisheries Lab of the Rhode Island Department of Environmental Management. Results of the scientific inquiries into the oyster farmer observations eventually led to the initial opening of a permanent breachway in Point Judith Pond during the early 1900s to aid in water quality issues that were impeding the oyster production there. This permanent opening also led to the expansion of the major fishing port at the village of Galilee.

While the Rhode Island aquaculturists fought for marine science and water quality improvements, the rest of the United States fell victim to industrial pollution and continued overharvest practices. For the cynical reader who finds it hard to believe that all oysters in the United States were harvested to near extinction, we can look again at Ingersoll's *The Oyster Industry* report for examples of the human capability of overharvest. Ingersoll's vivid description of the method of fishing the Connecticut oyster beds in the early days shows the inefficiencies of a "regulated" oyster harvest and how the harvest of one day could decimate an entire oyster bed. Ingersoll says:

> *The law was "off" on the 1st day of November, and all the natural beds in the State became open to any person who wished to rake them. In anticipation of this date, great preparations were made in the towns along the shore, and even for twenty miles back from the seaside. Boats and rakes, and baskets and bags, were put in order.*
>
> *The day before, large numbers of wagons came toward the shore from the back country, bringing hundreds of men, with their utensils. Among these were not unfrequently seen boats, borne on the rigging of a hay-cart, ready to be launched on the expected morning. It was a time of great excitement, and nowhere greater than along the Quinepiac.*
>
> *On the day preceding, farmers flocked into Fair Haven from all the surrounding country, and brought with them boats and canoes of antique pattern and ruinous aspect. These rustics always met with a riotous welcome from the townboys, who hated rural competition. They were very likely to find their boats, if not carefully watched, stolen and hidden before they had chance to launch them, or even temporarily*

disabled. These things diversified the day and enlivened a community usually very peaceful, if not dull.

As midnight approached, men dressed in oilskin, and carrying oars, paddles, rakes, and tongs, collected all along the shore, where a crowd of women and children assembled to see the fun. Every sort of craft was prepared for action . . . and they lined the whole margin of the river and harbor on each side in thick array.

As the "witching hour" drew near, the men took their seats with much hilarity, and nerved their arms for a few moments' vigorous work. No eye could see the great face of the church-clock on the hill, but lanterns glimmered upon a hundred watch-dials, and then were set down, as only a coveted minute remained. There was a hush in the merriment along the shore, an instant's calm, and then the great bell struck a deep-toned peal. It was like an electric shock. Backs bent to oars, and paddles churned the water. From opposite banks, navies of boats leaped out and advanced towards one another through the darkness, as though bent upon mutual annihilation.

"The race was to the swift," and every stroke was the mightiest. Before the twelve blows upon the loud bell had ceased their reverberations, the oyster-beds had been reached, tongs were scraping the long-rested bottom, and the season's campaign upon the Quinepiac had begun. In a few hours the crowd upon some beds would be such that the boats were pressing close together. They were all compelled to move along as one, for none could resist the pressure of the multitude. The more thickly covered beds were quickly cleaned of their bivalves.

The boats were full, the wagons were full, and many had secured what they called their "winter's stock" before the day was done, and thousands of bushels of oysters were packed

*away under blankets of sea-weed in scores of cellars. . . . That
first day was the great day, and often crowds of spectators
gathered to witness the fun and the frequent quarrels or fights
that occurred in the pushing and crowding. By the next day
the rustic crowd had departed, but the oysters continued to be
sought. A week of this sort of attack, however, usually sufficed
so thoroughly to clean the bottom, that subsequent raking was
of small account.*[*]

---

* Ingersoll, *The Oyster Industry*.

THE COLLAPSE

AMERICANS' APPETITE FOR OYSTERS WAS MET with an unfortunate foe: industrialization.

At the production peak, from about 1880 to 1910, the United States produced as much as 160 million pounds of oyster meat (twenty-seven million bushels) per year, more than all other countries combined! In the late 1800s, New Yorkers ate on average six hundred oysters per year, per person. Today that number is just an average of three oysters per person. Most of the labor used to shuck those oysters was from children.

In the early 1900s, photographer Lewis Hine famously exposed child labor exploitations by traveling the U.S. and taking pictures of horrific conditions children were forced to work in. One of the more famous images was taken in Dunbar, Louisiana, where Hine met an eight-year-old oyster shucker named Rosy. Rosy worked daily, from 3 a.m. to 5 p.m., and told Hine that "the baby of the family will start shucking as soon as she can hold the knife." Amazingly, none of these children wore gloves while shucking.

Lewis Hine, *All these work in Peerless Oyster Co. Had to get the photo while bosses were at dinner, as they refused to permit children in the photos.*

Lewis Hine, *Rosy, an eight-year-old oyster shucker who works steady all day from about 3:00 A.M. to about 5 P.M. in Dunbar Cannery.*

Oysters consumed all facets of American culture. Oyster bars became speakeasy havens for political events. In 1860, even Abe Lincoln appeared in political cartoons slurping oysters. Oyster carts and canned oysters provided Americans with affordable, nutritious protein.

Unloading Oyster Luggers, Baltimore, Md.

In 1909, oysters cost half the price of beef per pound. Oysters were eaten at breakfast, lunch, and dinner, rich and poor alike. People even owned special oyster plates and forks for serving and eating oysters, shaped and painted like oyster shells. The now-famous Chinese takeout box was originally patented in 1894 by Frederick Wilcox as an "oyster pail," a single piece of paper folded in such a way that it was leakproof and capable of carrying shucked oyster meats home from the street cart, an improvement over the traditional wooden receptacle.

Currier and Ives, *Honest Abe Taking Them on the Half Shell*

The consumption of oysters was immensely trendy. As one nineteenth-century author raved, "The oyster, when eaten moderately, is, without contradiction, a wholesome food, and one of the greatest delicacies in the world. It contains much nutritive substance, which is very digestive, and produces a peculiar charm and an inexplicable pleasure. After having eaten oysters we feel joyous, light, and agreeable—yes, one might say, fabulously well."[*]

All good things must come to an end. The turn of the century saw a dramatic collapse for the American oyster industry. Industrial factories sprung up along the coast, injecting pollutants into the coastal waters without regulation. The early 1900s saw breakouts of typhoid and gastrointestinal disorders connected

---

[*] Eustace Clare Grenville Murray, *The Oyster: Where, How and When to Find, Breed, Cook and Eat It* (London: Trubner & Co., 1861), 13.

to oyster consumption and industrial pollution. As water quality around the country began to deteriorate, oyster beds were closed for the sake of public health, and the general public became wary of eating oysters from polluted waters.

The 1921 annual report by the Rhode Island Commissioners of Shell Fisheries stated, "The main cause for our decline of our oyster industry is large quantities of oil floating on the waters of our rivers, bay and tributaries; but all authorities agree that the instant spawn come in contact with any oil, it is instantly killed."[*]

In 1927, the last of the New York oyster beds were closed due to overharvesting and pollution. A similar story took place across the whole of Europe, Japan, and elsewhere, as industrialized coastlines spewed chemicals into coastal waterways. The rise of global industrialization was the catalyst for the global collapse of the last remaining oyster industries. Much like the canary in the coal mine, oyster reef collapse is the telltale sign of polluted waters.

Fifty years of unregulated industrial pollution went unchecked before action was taken. During this time, oyster consumption fell out of popularity. Entire generations grew up avoiding oyster consumption, and with good reason; any remaining oysters were certainly growing in contaminated waters.

It wasn't until 1972 that America passed the Clean Water Act, a federal law that established the basic structure for regulating the discharging of pollutants into the waters of the United States. This was one of the first and most influential modern environmental laws in American history, regulating industrial pollutants and protecting coastal waters over corporate interests. The law also inspired international action.

---

[*] *Annual Report of the Commissioners of Shell Fisheries* (Providence: The Oxford Press, 1921), 10.

In 1984, China passed the WPPCL (Water Pollution Prevention and Control Law), their first legislation on pollution prevention with a focus on industrial pollution. The law has undergone several revisions, specifically in 1996 and again in 2008, as a result of severe water pollution across China. In 2022, China passed the New Pollutant Control Action Plan, its most focused plan to date, which looks at regulating "new pollutants" such as persistent organic pollutants, endocrine disruptors, antibiotics, and microplastics.

Europe passed a similar law in 2000 that requires all EU member states to achieve good status in all bodies of surface water and groundwater by 2027. In 2012, the EU passed the Marine Strategy Framework Directive, which aims to protect the marine environment. It requires the application of an ecosystem-based approach to the management of human activities, enabling a sustainable use of marine goods and services.

It has taken decades, but we are starting to finally see a positive change in water quality across the globe. Coastal pollution is now heavily regulated and state governments now classify ocean waters into various water quality grades, with the cleanest grades allowing for the taking of seafood. Water quality monitoring programs ensure cleanliness is maintained, while stringent checks and balances ensure harvested product remains safe and transparent. The improvements in water quality have allowed the oyster industry to rebound—in the form of aquaculture. The vast majority of oysters consumed in the world now come from farms. In the United States, harvested oysters are now the second most regulated food crop behind raw milk.

It is now estimated that 99 percent of the historical global wild oyster stock has been consumed or destroyed from overfishing and ecosystem degradation. Once the most numerous life forms on Earth, wild oysters should now technically be placed on the endangered species list. However, the tables are turning. With the protection of our coastal waters through environmental laws, shellfish restoration efforts, harvest regulations, and the creation of shellfish farming, we are finally starting to see a rebound in the global wild shellfish numbers.

While great strides have been made over the last few decades, the reality is that the ocean's wild fisheries can no longer sustain the growing human population's demands for seafood.

The global population has more than quadrupled in the last one hundred years, with an estimated eighty-two million people now added to the planet every year. By 2100, the estimated human population will be around eleven billion people. We will need to produce more food in the next fifty years than has been produced in the last twelve thousand years.

A recent four-year study of 7,800 marine species concluded that if overfishing does not stop, or if we fail to suspend the unsustainable fishing practices, the world could run out of seafood as we know it by the year 2048. This nightmare scenario is compounded by the fact that the UN Food and Agriculture Organization said in 2014 that at the current rate of soil loss and degradation, we have just sixty years of harvests left in the ground.

Ocean farming has the ability to meet these increased food production demands, while also bettering the ecosystems and environment. Like land agriculture, aquaculture will significantly transform our society, and it is already beginning to do so. Shellfish farming (clams, scallops, mussels, and oysters) produces human food and livestock feed without the use of fresh water, land, antibiotics, or fertilizers, and leaves the wild shellfish stocks intact to regrow and revitalize the natural ecosystems. Shellfish farming also combats ocean acidification, sequesters carbon, and bolsters a wide range of marine nursery species.

Over twelve different species of oysters are now farmed across the globe, and the oyster aquaculture industry has been growing at roughly 5 percent a year since the 1990s. Demand for oysters is now the highest it has been in decades, and with good reason; clean water means clean oysters.

As the shellfish industry adapts to the various coastlines around the globe, new methods, gear types, and farm systems are being created, all in the spirit of raising the best oysters possible while using the least resource-intensive footprint. Shellfish farming is now the most sustainable form of protein farming on the planet.

Technology and innovation breakthroughs are not only improving shellfish farms but are opening the floodgates for the aquaculture of multiple new oceanic species. "3D farms" are now pushing the limits of productivity by utilizing an X, Y, and vertical Z axis for farming, where different species grow side by side in an engineered ecosystem farm space.

New start-ups have sought to sequester carbon and mitigate nutrient buildup within coastal embayments, use robots for coastal monitoring and food harvest duties, or even attempt to grow land-based foods in an underwater greenhouse-like structure.

Investors are dumping enormous amounts of money into the "blue economy," and prospective farmers and governments are using these funds to zone and develop the coastlines into productive acres. We are now at the dawn of the next agricultural pillar for human society to stand on.

There is no future without farming the ocean.

THE PRESENT

CURRENT STATUS
OF THE INDUSTRY

I'VE SPENT THE LAST DECADE LEARNING TO FARM
the ocean. As someone who had zero background in boats or farming, I knew the only certainty was that I didn't know anything. I did, however, know that my best chance at success would be to learn as much about aquaculture farming as possible. My brother shared the same mindset, and so we began our shellfish farming journey.

From navigating the initial stages of ocean-farming permitting and applications, to studying the various types of farms, species, methods, and gear, to building oyster farm boats and acquiring coastguard credentials, learning the biology and shellfish survival needs, to studying market research opportunities, experimenting with new oceanic species and building brand awareness, I've tried to absorb as much about the industry, trends, and common practices as possible. Much of our success has been built on the shoulders of those who came before us. We have also been able to recognize the direction in which the industry and market is moving and have tried to put ourselves in front of it while the opportunities have presented themselves.

The oyster industry has made great strides in the last decade to become the fastest growing and one of the most exciting new agriculture industries to be a part of. Around the world, oyster restaurants have seemingly sprung up overnight and oyster consumption has become trendy again. In response to increased consumer demand, a modern-day gold rush is taking place, as prospective farmers stake their claims and make a go of oyster farming.

Global aquaculture production (all species) has grown significantly over the past fifty years and now accounts for more than 50 percent of the world's seafood supply. Mollusks (shellfish) are the second-largest category of farmed seafood by both quantity and value, accounting for 21 percent of all global aquaculture production by weight in 2016. The leading molluscan species by quantity produced is oysters.

In 2018 (pre-Covid), China was by far the largest producer of farmed oysters with 5.14 million tonnes, or 85 percent of the global market share. Amazingly, China consumes almost all of its own production. Followed by China is Korea with just 303,200 tonnes of oysters, Japan with 176,000 tonnes, and the U.S. with 153,909 tonnes.

While these numbers may seem huge, they make up only a fraction of the historical oyster consumption totals. Remember that in our not-too-distant past, Earth (with a much smaller human population) consumed almost six hundred times more oysters than current consumption rates.

Astonishingly, only about fifty thousand tonnes of oysters are traded internationally. The majority of exports originate in Korea and France. Less than 5 percent of the global oyster production enters international markets—one of the lowest proportions in seafood trade. This is largely due to the nature of the oyster, being a highly perishable item with the potential risk to human health if not handled correctly. Put another way, 95 percent of oyster production is consumed within the producing country, which makes the oyster one of the most locally sourced foods eaten. With demand forecast to continue to rise for decades, international trade may never take off, except for specialty high-value products.

As previously discussed, the oyster industry has generally not been stable at any location. Geographic events, disease, parasitic outbreaks, water quality issues, and overharvesting are typically to blame for industry collapse.

Although many countries have had difficulty maintaining long-term oyster aquaculture industries, production on a global scale has rapidly increased since the 1950s (the beginning of data collection). In 1990, 1.2 million tonnes of oysters were produced globally. That number jumped to 6 million tonnes by 2018, with a global value of $7.46 billion. A five-fold increase! The industry is expected to continue growing at a compound annual growth rate of 3 to 5 percent for at least the next decade or two.

The rapid increase in shellfish aquaculture growth can be attributed to water quality improvements, recent innovations within the industry, and a newfound awakening in cultural awareness of agriculture's impact on the planet. As more people seek proteins that are less resource-intensive to produce, shellfish aquaculture rises to the forefront as the most sustainable form of protein farming on the planet.

This increase in demand has brought much-needed development capital. Funding is flooding into the industry and technological breakthroughs are enabling access to new species and additional sources of seed.

A variety of oyster species are now available to be farmed across the globe thanks in part to shared research and improved hatchery capabilities. Hatchery technology has reduced the reliance on natural seed sources for farm operations and has introduced more productive strains to areas impacted most by wild stock collapse.

Advances in gear designs and materials have improved efficiency, productivity, and quality in shellfish products. Shellfish farms inherently utilize minimal to zero freshwater or feed inputs, and improved designs now allow off-grid "green energy" farms to exist in some of the most remote areas of the world. Improved farm methods and harvest practices deliver safe and clean products with more transparency than ever.

One of the most exciting aspects of the oyster industry is that, like wine, the regionality, species farmed, farm techniques, and water varieties that go into each oyster create an enormous amount of difference in the product, and just like with wine, these differences are celebrated.

In the wine world, the environmental influences on the product create the "terroir"—the complete natural environment in which a particular wine is produced, including factors such as the soil, topography, and climate. Terroir in its simplest form is "the taste of the land." It's the entire combination of all the terrestrial influences that went into the bottle of wine.

In the oyster world, the French have created a word for these environmental variations: merroir. The taste of the sea. It's the combination of all the marine influences that flavor the oyster, such as the species type, bottom sediment type, salinity, and climate.

Before we dive into the aspects of the oyster industry that separate the good from the bad, the clean from the dirty, and the expensive from the cheap, it would make sense to start at the beginning of the modern aquaculture farming process.

HOW TO
ACQUIRE A FARM

THE PROCESS TO OBTAIN AN OCEAN FARM PER-
mit can be a daunting task. Rules and regulations typically vary greatly between towns, states, and countries.
There is no set of uniform guidelines. Some communities embrace
the notion of ocean farming and tend to support the applicant on
their endeavor, while other locations will literally chase the applicant out of town.

First and foremost, comes ownership. In the United States, a
state owns the water; however, the municipality has the authority
to zone the water (permit water use activities) as it sees fit. European countries act in a similar manner, allowing towns the final say
in permitted marine activities.

States own the ocean from the high tide line to about twelve
nautical miles out to sea. The federal government owns the waters
from twelve nautical miles to about two hundred nautical miles

out from shore. European countries follow this general rule as well. After two hundred nautical miles, the ocean becomes international and is considered the "high seas." The high seas make up 50 percent of the surface area of the planet and cover over two-thirds of the ocean.

When the first colonists came to the New World, they brought with them the European ideals that the ocean was a public common, owned by the state but available for public use. This mentality has carried forward, and ocean farmers in New England must apply for and lease the ocean bottom from the town and the state. Most towns have legislation that requires the farmer to be a town resident.

American East Coast farms are therefore not private property, and recreational activities of all sorts may be carried out within their bounds. However, once deeded a farm, only the farm entity is allowed to harvest from the area. Commercial fishermen are typically not allowed within the farm zone. More on this later.

Developed during the Gold Rush "stake your claim" era, the American West Coast ocean can be directly owned by an individual, who pays for the lease deed just like a land title. West Coast farms are generally considered private property, and recreational activities are not permitted within their bounds.

When applying for a farm on either American coast, the individual must first present their application through the local town authorities. This may be a board, mayor, shellfish committee, conservation commission, or some other kind of local entity authorized to permit ocean activities. Each town has a different set of rules for what type of aquaculture activities are allowed and who is permitted to do them. For instance, my local town (Oak Bluffs) requires that shellfish farm applicants be year-round town residents and meet the minimum state requirements for a commercial fishing license.

In addition to these requirements, my town does not allow aquaculture farms within the coastal salt ponds and does not allow floating aquaculture growing gear. The neighboring town allows both of these practices. When I applied for my farm license, I had to formulate my proposal to meet these requirements, and upon a preliminary approval by the local town entities, the state was allowed to step in.

Each state process is different. In Massachusetts, our farm proposal was sent to the Army Corp of Engineers, Coast Guard, the Native American Wampanoag Tribe, Massachusetts Department of Environmental Protection, Conservation Commission, Board of Underwater Archaeological Resources, Division of Marine Fisheries, and Division of Fisheries and Wildlife. Seems a little excessive for an ocean farm, doesn't it? Land farms in comparison have little to no government oversight when being formed.

Each department reviewed our proposal in respect to their purview and weighed in with a letter of response to the Oak Bluffs selectmen. Within three months all responses were returned in favor of our project. The last department to weigh in was the Massachusetts Division of Marine Fisheries (DMF). A scuba dive survey of the area needed to be performed before they could sign off on the farm location. This service is typically provided by the state.

Like most state fishery management agencies, the DMF will only grant permission for a new farm site if it meets strict criteria and does not interfere with existing commercial fishing or established ecosystems. The states do not want to replace fishermen with farmers, they want to add farmers to areas where fishermen do not work. In this way, the state can maintain the existing commercial fisheries while also growing the new aquaculture industry. It is a smart approach to what could be a complex problem.

During our application process, a team of DMF divers were

dispatched onto our proposed site and they systematically swam in a grid formation, counting any existing commercial species in the area. In addition to looking at commercial fishermen catch reports and commercial species present on the site, they also looked for existing eelgrass (a type of seagrass) or ecological habitats. If any were discovered, our application would have been denied.

Luckily, our open-ocean site contained a minimum number of wild shellfish species and no eelgrass beds. Referencing existing catch data, the DMF was able to deduce that little to no commercial fishing took place in our proposed area. User conflict would be minimal, and they granted us permission to move forward.

For a lot of prospective farmers, the application process is considered complete after the state signs off. A farm permit is granted by the town for X number of years (according to their local regulations) and the state issues a propagation permit to the farmer to purchase and raise juvenile shellfish from an approved hatchery. The applicant pays their annual fees to the town and state and starts to throw a chunk of investment money into the ocean, hoping it comes back as a small return a few years later.

For a lot of other prospective farmers, approval by the local town only brings about lawsuits by abutting property owners (as happened in our case). These lawsuits are typically groundless and are usually a case of NIMBY-ism (Not In My Back Yard). The major scare for these individuals is that the farm will somehow diminish their property values.

There is no example to be found of a shellfish farm diminishing the value of a neighboring property. In our case, as with most others, shellfish aquaculture farms not only provide a benefit to the surrounding ecosystem and water quality, but they also create a strong economic engine for the local economy. Jobs are created by the farm, money is spent by the farm, and most importantly, the shellfish products create value for the local restaurants and suppliers. Shellfish aquaculture is typically assigned a 3x economic multiplier effect. For every dollar generated on the farm, it generates another three dollars into the local community!

The lawsuit against our farm approval warranted additional public hearings. As these drew on it became clear that the town would defend its approval of our application and was willing to take the plaintiffs to court if necessary. In reality, the plaintiffs had no leg to stand on, arguing that the process had been rushed and the town had made an ill-informed decision. To avoid a lawsuit and a potential cease and desist, we scheduled a special meeting with the neighbors to listen to their concerns, which is what they ultimately wanted—to be heard. It pays to be a good neighbor and listen. The lawsuit was dropped and we were able to start the journey of ocean farming. It took nearly two and a half years from the day we filed our application to the day we were allowed to start farming.

Unfortunately, not every application ends with a happy farmer. The number one reason for lawsuits and negative pushback is because of the lack of education on shellfish farming. There is not enough material readily available about the industry, although this book hopes to change that.

Once the farm is permitted, the real work can begin.

DIFFERENT TYPES
OF OYSTERS

THERE ARE MORE THAN TWO HUNDRED DIFFER-ent species of oysters around the world. Several species of oysters are farmed for pearls; however, just thirteen species are farmed globally for human consumption. Of the thirteen, only five are available on a large scale—though only within their local regions. Like wine grapes, each oyster variety has classic physical and taste characteristics, which can be expressed differently depending on location and farm conditions used.

Here is a quick breakdown of the thirteen farmed oyster species for human consumption, ranked in order of global production value:

- **Pacific oysters (*Crassostrea gigas*):** Farmed around the world. Predominantly grown in China, Japan, Korea, Taiwan, Thailand, Australia, New Zealand, Malaysia, Canada, the West Coast of the U.S., Mexico, France, Ireland, the Netherlands, the Channel Islands, the U.K., Spain, and Portugal.

Large, soft, sweet, with cucumber notes. Sometimes referred to as "Rock oysters" in Ireland and Britain.

- **Atlantic oysters (*Crassostrea virginica*):** Grown predominantly on the East Coast of the U.S., Canada, and the Dominican Republic. Large, firm, briny, and sweet.
- **Belon/European flat oysters (*Ostrea edulis*):** Predominantly grown in France, Ireland, the Netherlands, the Channel Islands, the U.K., Spain, Croatia, and Portugal. Metallic notes, with a trace of caviar and hazelnut. The original flat oyster of Europe.
- **Kumamoto oysters (*Crassostrea sikamea*):** Grown on the West Coast of the U.S. and in Japan, Korea, and Taiwan. Small, creamy, with hints of melon.
- **Sydney rock oysters (*Saccostrea glomerata*):** Grown in Australia and New Zealand. Known as "Rock oysters." Small, creamy, minerally.
- **Olympia oysters (*Ostrea lurida* or *Ostrea conchaphila*):** Native and grown on the West Coast of the U.S. Tiny, coppery, smoky.
- **Chilean flat oyster (*Ostrea chilensis*):** Grown in Chile and New Zealand. Related to the European flat oyster. Also known as *ostra verde*, the Bluff oyster, or the Dredge oyster. Flat oyster. Metallic and sweet.
- **Slipper-cupped oysters (*Crassostrea iredalei*):** Grown in the Philippines. Brackish water species. Creamy flesh.
- **Hooded oysters (*Saccostrea cucullata*):** Grown in Mauritius, East Africa, and the Indo-Pacific on mangroves. Also known as the Natal rock oyster. Silky.
- **Cortez oyster (*Crassostrea corteziensis*):** Grown in Mexico and Chile. Mangrove oyster. Little is known about this oyster.
- **Mangrove cupped oyster (*Crassostrea rhizophorae*):** Grown in Cuba, Puerto Rico, Brazil, and Jamaica. Intertidal species.
- **Gasar cupped oysters (*Crassostrea gasar*):** Grown in Brazil, Senegal, and Gambia. Mangrove oyster. Little is known about this oyster.
- **Indian backwater oyster (*Crassostrea madrasensis*):** A brackish water oyster found along estuaries of the southwest and east coasts of India.

HATCHERIES VS.
WILD SPAT COLLECTION

REGARDLESS OF WHICH OYSTER SPECIES THE farmer chooses to grow, there are only two ways to acquire baby oyster seed. The most tried and true way is to collect the wild baby oyster spat similar to how Sergius Orata did it over two thousand years ago.

Spat collectors are distributed along the waterway with the hope that wild oyster spat will float along and attach to the collector. These collectors come in various forms, from floating sticks like the Orata method to concrete plated discs or posts. Luckily, the oyster spat is not too picky. The trick is planting the collectors in the environment at the right time. Too early, and the collectors can become fouled with other organisms. Too late, and the natural oyster set can be missed entirely. Once the oyster has attached and is growing on the collector, the farmer can then separate the oyster from the collector and place it into the farm growing system.

The largest challenge with wild spat collection is that the success rate is highly variable. One season may produce a lot of seed while another season may miss the wild spat completely. Individual oysters are hard to come by when collecting wild spat. Oysters naturally settle on top of each other, creating clumps and reefs. When large amounts of spat are collected, the oysters can become irregular, variably sized, and differently shaped. They tend to look "wild." Much of the global supply of Pacific oyster seed is obtained from wild seed capture.

The second way of obtaining baby oyster seed is by purchasing it from a hatchery. There are only a handful of hatcheries in the United States, yet they form the foundation for the entire aquaculture industry. The vast majority of oyster farms in the U.S. acquire their seed from a hatchery.

A shellfish hatchery is typically a building located along a shoreline that has an intake pipe slurping water from the nearby waterbody into the building. The hatchery uses this natural water, filtered and purified, to spawn and grow oysters until they are large enough to be transferred to the farmer.

The spawning process is not easy. For starters, the hatcheries must filter and balance the pH of the incoming water. Fouling organisms are removed while any pH oddities are balanced. This wasn't always the case.

In 2005, production at shellfish hatcheries along the American West Coast began to decline. Just three years later in 2008, 80 percent mortality was experienced by the largest hatchery on the West Coast, and nobody understood why. Farms collapsed without the hatchery seed. Only the largest farms, which had the capacity to collect wild spat, survived.

The problem was not with the oysters. Hatchery biologists were able to spawn the parent oysters like normal, and they could see the larvae floating around on the surface waters. They watched as the larvae settled on their tiny grains of crushed oyster shell. Metamorphosis growth was normal. But when the oysters began to grow shells, they became deformed, weak, and brittle. Eventually the baby oysters would die.

It wasn't until 2010, five years after the first mortality rates, that biologists finally determined what was responsible for the collapse of one of the largest oyster industries in the world: ocean acidification. It turned out that the waters along the West Coast had grown so acidic that it was actually dissolving the oyster shells before the biologists' eyes.

According to the National Oceanic and Atmospheric Administration, "Ocean acidification refers to the ongoing change in the chemistry of the ocean caused primarily by the ocean's absorption of carbon dioxide from the atmosphere. The ocean generally has an alkaline pH, but as the ocean continues to absorb carbon dioxide it has become more acidic. As the ocean becomes more acidic, carbonite-based minerals such as calcium carbonate, important for shellfish development and survival, are reduced."*

To combat the acidifying waters, hatcheries have now installed pH monitoring stations to capture and correct the pH of the hatchery water prior to it being used in spawning activities. Once the oyster has grown to a large enough size, it can defend and repair itself against slightly more acidic ocean waters. However, as the ocean continues to grow more acidic due to carbon dioxide pollution, wild shellfish populations will lose the ability of their larvae to survive. It is already happening on the West Coast. Ocean acidification is the number one concern facing the

---

* "Pacific Ocean Acidification Project," NOAA, ioos.noaa.gov/project/pacific-ocean-acidification-project.

shellfish population in the future. If the ocean continues to grow more acidic, we could face a future in which shellfish cannot successfully replicate in the wild.

Luckily for the American East Coast hatcheries, the Atlantic Ocean is growing more acidic at a slower rate than the Pacific Ocean. East Coast hatcheries have learned from the West Coast dilemma and have installed pH monitoring stations at their facilities in preparation for future acidic waters.

WITH THE INCOMING hatchery water safe and sound, spawning of the oysters can begin, usually taking place in the cold winter months so that the babies are of a certain size by the time the warmer spring temperatures are present on the farm.

In order to spawn in the winter months, the water must be heated to 65 degrees Fahrenheit, which tricks the oysters into thinking it is summer. Contained within small aquariums, the male oysters release their sperm and the females release their eggs. Millions of eggs fertilize, metamorphosize, and quickly begin looking for food.

In order to support hundreds of millions of baby oysters, the hatcheries must also grow their own large cultures of phytoplankton to feed the newly spawned babies, as naturally occurring phytoplankton is typically not present in large enough quantities in the northern latitude wintertime. Like a coveted sourdough culture, some phytoplankton strains have been cultured for fifty-plus years and are protected like intellectual property. However, it's not so much the individual strains of phytoplankton that are protected, but more the concoction of strains fed to the oysters.

Shellfish eat by filtering and sifting through the seawater to extract specific strains of algae and nutrients needed for their health. Of all the shellfish species, oysters are the rock stars when it comes to filtering. Oysters can filter up to fifty gallons of water a day and have the finest gills of any shellfish, able to filter particles down to just one micron in size.

Oysters are classified as herbivores because they feast solely on phytoplankton, which are microscopic pieces of seaweed. Any other compounds, such as zooplankton or foreign materials that they ingest through filtration, are quickly expelled. Only the algae are consumed.

Phytoplankton are microorganisms that drift about in the water and create the foundation of the oceanic food web. They are single-celled but sometimes grow large enough to be seen by the human eye. Sometimes blooms of phytoplankton grow so large that they can be seen from outer space. Phytoplankton rely on photosynthesis to produce sugar for energy, but they still need other nutrients to grow and reproduce, typically phosphorus, nitrogen, and iron—some of the most plentiful nutrients on Earth. The more nutrients present in the environment, the more algae and phytoplankton will grow. Like a golf course pond, if there are too many nutrients, the algae will smother the waterbody, consume all available oxygen, reduce water quality and aquatic health, and can eventually kill all living organisms within the ecosystem. This is the situation that has been occurring across our waterways for the last few decades as unchecked nutrient runoff has flooded the coastal ponds and estuaries.

Phytoplankton are divided into two classes: algae and cyanobacteria. These two classes contain at least one form of chlorophyll (chlorophyll *a*) and can conduct photosynthesis for energy. They can be found in fresh water and salt water.

There are thousands of species of microalgae floating in water all over the world. Green algae, diatoms, and dinoflagellates are the most well-known. Diatoms have garnered the most attention of the microalgae species, having become a cultural phenomenon upon their discovery in the early 1800s.

Diatoms are microscopic single-cell algae that create an exoskeleton of beautiful glass shells. There are hundreds of thousands of varieties of diatoms, all with unique forms. Scientists who discovered these tiny organisms were enthralled. Utilizing a microscope and needle-like tools, the scientists would align the diatom shells into artistic arrangements on the microscope slide. These slides became the subjects of diatom parties, at which scientists and intellectuals would display their creations. The diatom art reached its height at the end of the century and is a great example of the Victorian desire to bring the world into order.

Today there are so many diatoms drifting in the oceans that their photosynthetic processes produce about 50 to 80 percent of the Earth's oxygen. Every other breath you take was created from ocean phytoplankton.

Because phytoplankton are able to produce their own energy through photosynthesis, they are considered autotrophic (self-feeding). Autotrophs are called primary producers because all other organisms rely on them (directly or indirectly) as a food source. Phytoplankton are generally consumed by zooplankton and other small marine organisms like krill and shellfish. These creatures are then consumed by larger marine organisms, up the chain to apex predators, including dolphins, sharks, and humans.

It has recently been discovered that all nutritious health benefits within ocean organisms originate from phytoplankton, which contain an extraordinary range of macronutrients, micronutrients, and phytonutrients essential for good health. Phytoplankton are around 50 percent protein by dry weight and just under 10 percent carbs. Omega-3 fats; vitamins A, C, E, K, and a wide range of B; iron; magnesium; calcium; potassium; phosphorus; trace elements of chromium, selenium, boron, iodine, and zinc; chloro-phyll; carotenoids; and phycobiliproteins are all contained within phytoplankton organisms.

This incredible range of nutrition is the sole source of food for the oyster. Day by day, the oyster sifts through the ocean's bounty with its delicate fingerlike appendages called cilia. The unwanted ocean detritus and debris are cast away, and only the best micro-scopic pieces of algae are selected, passed from cilia to cilia into the mouth of the oyster and consumed. They are then digested in the stomach, where the pure amino acids of the phytoplankton's health are morphed into the succulent oyster meat and flesh, and the remaining calcium and carbon are converted into shell. For a human, the oyster is nature's multivitamin. For an oyster, the phy-toplankton is nature's multivitamin.

Back in the hatchery, proprietary blends of phytoplankton are used to feed and grow the seed oysters until they reach a size large enough for the farmer to take. One millimeter in size (about the size of a pencil lead tip) is the smallest oyster a farmer can take. The price of the baby oyster varies by size, typically around one to two cents for a one-millimeter oyster, up to thirty-five cents for a one-inch oyster.

As the newly spawned oysters come to size and grow, the hatch-eries gradually lower the temperature of their water to come into an equilibrium with the warming outside waters in their respective attached waterbody. The goal is to have the hatchery water tem-perature equalized with the farm water temperature, and the oys-ters at the proper size, just as the spring algae bloom on the farm is beginning to flourish. Oyster seed is typically sold to the farmer at the beginning of the summer, end of May or early June. This gives the farmer an extra two or three months of growth compared to natural sets, which typically settle in mid- to late summer. In essence, the hatchery is like a greenhouse; it gives the farmer a head start on the growing season. In addition to providing farms with oyster seed ahead of the natural spawn, hatcheries are also responsible for the modern-day half shell market.

In the 1970s, American shellfish hatcheries discovered a process to eliminate the problem of oysters settling onto other oysters and creating undesirable clumps. For the first time in history, biologists at the Milford Lab in Connecticut measured the exact micron size of the oyster larva's foot and discovered that if they ground oyster shell into a matching-sized grain and spread these grains evenly across the bottom of an aquarium, the oyster larvae would individually settle onto a single grain of shell. Millions of oysters settle onto millions of grains of shell. This phenomenon created the ability to count, weigh, and farm individual oysters. When we buy five hundred thousand baby oysters for our farm, there are literally five hundred thousand individual grains of oyster shell attached to each oyster foot. After the larvae oysters settle onto the grains of shell, the oysters will go through a metamorphosis, shedding the eye spot and flipper. The foot remains cemented to the grain of shell for the rest of the oyster's life, and the oyster begins to form a calcium carbonate shell of its own.

Almost overnight the oyster farming industry went from having to collect wild clumps of oyster spat that needed to be ripped apart, to having baby oysters individually settle onto a grain of shell and be fed proprietary blends of phytoplankton to obtain maximum growth and health, while being delivered months before any wild spawns took place. The vast majority of oysters served on the half shell today once settled onto a single grain of shell in a hatchery.

Unfortunately, there are not enough hatcheries in the United States to keep up with farm demand. Seed must be ordered a year in advance to guarantee delivery and even then, delivery of seed is not guaranteed. A general rule is to order seed from multiple hatcheries, each with parent stock that comes from a more northerly climate than the farmer's growing area. For instance, all seed grown on Martha's Vineyard comes from hatcheries in Maine. The northern Maine genetics flourish in the milder Vineyard winters. If we bring seed from Texas to the Vineyard, it would surely die, having never spent a winter in freezing temperatures.

Baby oysters being planted on the Cottage City oyster farm

DIPLOID VS. TRIPLOID

WHEN THE OYSTER FARMER PURCHASES oysters from a hatchery, there are typically two very different varieties the farmer can choose from: a diploid or a triploid oyster.

An oyster with two sets of chromosomes is called a diploid. Diploid oysters are the naturally occurring oysters we find in the wild. They grow, mature, spawn, and continue the cycle. Most animals in nature have two sets of chromosomes, one from each parent. Chromosomes are tiny structures of DNA made up of many genes, and in oysters, these genes determine traits such as shell height, cold tolerance, or resistance to certain diseases. Crossbreeding diploid oysters over multiple generations has been successful for breeding advantages or traits; however, the diploid oyster remains fertile and will therefore spend much of its life and energy dedicated to creating gonads to produce sperm or eggs. From the perspective of a farmer, who may not be interested in breeding oysters, this reproduction "waste" of energy could be thought of as a negative trait.

Triploid oysters solve this dilemma. Rarely occurring naturally in the wild (though they sometimes do), triploid oysters have three sets of chromosomes and do not have the ability to reproduce. Much like seedless watermelons, the benefit of triploid oysters is that they are sterile and will typically grow faster and meatier than their diploid counterparts. Triploids expend zero energy on reproduction and therefore remain "fat" all year long.

Triploid oysters are a relatively new variety for the oyster farmer, having been invented in 1979 by (at the time) grad student Standish Allen Jr.

Allen and other grad students of the Darling Marine Center, overlooking the Damariscotta River, were using Sea Grant funding to help create an aquaculture industry in Maine. Allen was learning the science of aquaculture, taking anything that worked with one species and testing it with others. He adapted a Norwegian technique that had been used on salmon, and tried forcing extra chromosomes into *Crassostrea virginica* oyster eggs, using a toxic chemical called cytochalasin B. The chemical never worked with the salmon but proved to be one of the keys to inventing the triploid oyster.

Allen learned that the chemical had to be applied during the window of time when the oyster egg met with the oyster sperm. During this short window of time, the fertilized egg throws off a set of female chromosomes as it takes up a set of male chromosomes. When Allen added cytochalasin B during this pivotal moment, the egg would keep both sets of female chromosomes while also adding a set of male chromosomes. The triploid oyster was born.

The fledgling Maine industry had zero interest in farming triploid oysters, and Allen's invention seemed like a flop. He departed for his PhD work at the University of Washington on the West Coast and soon found a thriving oyster industry excited by his triploid prospects. With support from Washington Sea Grant, Allen teamed up with Sandra Downing, another grad student, to apply the cytochalasin B treatment to a new species of oyster, *Crassostrea gigas*—the Pacific oyster. Working with the Coast Oyster Company, Allen and Downing were able to create successful commercial-sized batches of triploids with a 90 percent success rate. The West Coast farms quickly adapted the triploid species, and large hatcheries began selling the seed stock to farms along the coast.

Interestingly, triploid oysters are not considered genetically modified organisms (GMOs) because they are just a manipulation of the genetics, as opposed to a modification. The term GMO typically denotes a modification that cannot take place in nature, such as gene splicing. While triploids rarely happen naturally, they still do naturally occur, and the invention of the triploid oyster has been labeled a success of modern-day biotech.

The triploid process inspired another young man named Ximing Guo, who left China for the University of Washington, where he read Standish Allen's work on triploids and decided that for his PhD project he would create another new species of oyster: the tetraploid (with four sets of chromosomes).

Guo knew that tetraploid oysters would be the best way to create triploid oysters. A tetraploid oyster with four sets of chromosomes could mate with a diploid, and the offspring would create a triploid oyster with three sets of chromosomes. Tetraploid oysters would remove the need for cytochalasin B, which was considered a major carcinogen and which the Food and Drug Administration was getting ready to ban.

Guo's work in the area was met with unrelenting failure. He soon discovered that all diploid eggs were too small to contain four sets of chromosomes. His belief was that only a triploid egg would be large enough to hold four sets of chromosomes. But how would he find an egg in a creature that was created not to breed?

In 1993, Allen helped recruit Guo to Haskin Shellfish Research Laboratory in New Jersey. Allen was familiar with Guo's work and his hypothesis about triploid eggs. Interestingly, in Allen's early days of triploid creation, he would occasionally come across a triploid oyster with eggs. Much like a seedless watermelon with seeds, although rare, it can happen.

Allen and Guo recruited the aid of grad students and lab workers to meticulously slice open thousands of triploid oysters and examine them under a dissecting microscope for the exceedingly rare eggs, and eventually some were found. Working with cytochalasin B, Allen and Guo successfully implanted the triploid eggs with a fourth set of chromosomes and grew their altered eggs into oyster babies. The tetraploid oyster was born.

Guo was correct in his hypothesis. The tetraploid oyster was patented, and the technique is now used today to create triploid oysters for farming around the world.

In 1997, Allen led a campaign to introduce the non-native Pacific *Crassostrea gigas* and Chinese *Crassostrea ariakensis* oysters into the Chesapeake region. Triploid stock was a requirement of the experiment, and Allen successfully created the *Crassostrea ariakensis* triploid for the project, his third successful triploid oyster species.

Growth trials of both species were handsomely successful, but consumers claimed the Pacific oyster was too metallic-tasting for consumption. The Chinese *ariakensis* oyster, however, fared well with consumers.

After five years and millions of dollars spent, the research project was handed a death blow in 2009 when multiple state agencies and the Army Corps of Engineers released an environmental impact statement citing that the Chinese oyster might accumulate viruses harmful to humans and may outcompete the native *Crassostrea virginica* species. In the ruling, non-native oyster species such as the Chinese *Crassostrea ariakensis* variety would be prohibited from growing in Chesapeake waters.

Although the research project was a failure in introducing new aquaculture species to the region, it did foster the rapid acceleration of aquaculture practices in the Chesapeake Bay.

During the large-scale field trials, growers were asked to grow native *Crassostrea virginica* triploid species side by side with the Chinese counterpart as a baseline comparison. Those growers became the early adopters of the native triploid variety, which outgrows diseases in the area and creates a fatter product that has high demand in the summer months.

Today nearly 100 percent of all the seed stock used in the Chesapeake Bay is triploid, almost 100 percent of all French oysters are triploid, 15 percent of Sydney rock oysters (*Saccostrea glomerata*) in Australia are triploid, and roughly 50 percent of all West Coast oysters are of the triploid variety. While the growth may be better and the meats may be meatier, I and other "purist" farmers are against the use of triploids. It's not a health or safety issue, or a GMO or taste issue; it's a forward-thinking issue—instead of viewing the spawning of babies as a negative trait, we view this as a major positive for the future.

One of the benefits of farming traditional diploid oysters is that they spawn, and these babies drift into the oceans and help repopulate the coastline oyster reefs that have been overexploited for decades, and if these reefs can recover, they have the ability to restart the ecosystems that have been depleted in their absence. Who cares if the meats become a little watery after the oysters have spawned? The tradeoff is that we have created a product that, if

allowed to spawn and rebuild the wild population, has the ability to repair our ocean ecosystems. We've put action behind this
thinking. On the Cottage City Oysters farm tour, we encourage all
guests to discard their shells into the waters surrounding our farm,
so that when our oysters spawn, those babies will find the shells
and begin to rebuild the reef. Slowly but surely, we are building the
foundation for an improved marine ecosystem.

NURSERY SYSTEMS

REGARDLESS OF WHICH VARIETY OR SPECIES the farmer chooses, the size of the seed purchased and the nursery method of growing it are choices that can make or break a farm.

The smallest seed size that is available for purchase from a hatchery is one millimeter. At this size, the oysters are barely bigger than a grain of sand. Unable to count them individually, the hatcheries first count out a small amount of seed and then estimate the larger count of the seed by volume.

Five hundred thousand oysters at one millimeter in size, the quantity and size our farm purchases annually, costs about .01 cents each and arrives to us from the hatchery via FedEx in a Styrofoam cooler with icepacks to help keep the seed dormant during shipping. Five hundred thousand oysters take up a space roughly the size of a basketball.

Interestingly, there is an additional 7 percent Rutgers University fee on the sale of every *Crassostrea virginica* oyster seed created. When inquiring about the fee, I learned that in the early 1960s, two oyster diseases (Dermo and MSX) spread across the American East Coast, killing an estimated 90 to 95 percent of the wild oysters growing within the region. A team of Rutgers scientists collected some of the surviving oyster populations that exhibited natural immunity to the diseases and began a breeding program using the immune stock. All hatchery oysters today come from the descendants of those disease-resistant oysters, justifying the 7 percent kickback to Rutgers upon the sale of every oyster from a hatchery. Talk about a perpetual endowment!

The most important requirement of hatchery seed is that the hatchery is listed on a farm's state-approved hatchery list. In order for a hatchery to be approved to the list, the hatchery must provide a pathology report to the state for all seed being sold in state. The pathology report is a diagnostic report of tissue from the hatchery oysters to confirm the seed is free of disease or pathogens. Clean seed helps to ensure that wild populations of shellfish surrounding a farm are not subjected to disease imported by the farmer's seed. In a worst-case scenario, the farmer imports seed with diseases and wipes out the entire wild population in their region. A pathology report ensures this scenario will not happen.

Some farmers opt to only purchase seed that is already a year old and roughly one inch in size. While seed this size costs almost thirty-five times more than the smallest seed sizes, it allows the farmer to skip the hassle and skill of trying to grow smaller sized seed. This is a great strategy for a new farm and farmer who is likely limited by funds and experience in the location and who is hoping to reduce the amount of time needed to bring product to market. As farmers and farms mature, they typically desire to improve margins by purchasing smaller seed.

The growth of small oyster seed (one millimeter to one inch) is considered "intermediate growth" and requires a nursery system. There are some farm operations who *only* specialize in intermediate growth.

The first major distinction between nursery systems is whether they are land-based or located in the water.

Land-based nursery systems typically consist of a holding tank that the oysters sit in while water from a nearby water source is pumped in. This water contains all of the naturally occurring elements the oysters need for growth, such as oxygen, food, and shell-building materials. If the water is pumped up through the oysters, coming from beneath them and then going up and out, the system is called an "upweller." If the water is pumped onto the oysters like a waterfall, this is called a "downweller." If the water enters one side of the nursery container and exits the other side horizontally, the system is called a "raceway." To each their own. Some folks swear by one method over the other.

Regardless of the system, the method is the same. Super-small oysters are placed on a mesh, where they eat the naturally occurring algae in the water and grow. Over time, the mesh must be kept clean, and eventually, as the oysters get bigger, the mesh size must get bigger to allow more water flow, which provides more food. A good rule of thumb is to always have the mesh size be half the size of the oysters. For example, one-millimeter oysters start on five-hundred-micron mesh. One-inch oysters grow on half-inch mesh.

Each farm uses their own specific size and shape of container to hold the oysters. Each farmer cleans their oyster mesh a different way. Each farmer graduates their oysters to a larger mesh size a different way and at a different time. This is where the nursery location plays a huge role.

Some waterbodies are more productive than others. Some

oysters will only need to spend a few weeks in a nursery before they reach one inch and become large enough to plant on the farm. Other waterbodies are less productive and the oysters need months in a nursery system before graduating. Fun fact: The industry is still so small that most farmers still hand build and create their own nursery systems. There is no readily available one-size-fits-all, off-the-shelf nursery system that will work. *Every* location is different.

Water-based nursery systems work on the same principles as their land counterparts, except that the nursery containers are not on land, but instead sit in a floating raft. Usually, this raft is tied to a dock on the farm property. An electrical pump is used to either upwell, downwell, or horizontally pump water through the raceway. The benefit of a water-based nursery is that the oysters are growing *in* the water! The pump is not necessary, but used simply to increase the water flow through the system, which provides the oysters with an abundance of nutrients to promote accelerated growth.

As an industry that uses little to no fresh water or fertilizers/feed inputs of any kind, the oyster industry prides itself on being sustainable, eco-friendly, and organic by default. This is one of the reasons the industry is so attractive. In this spirit, leaders in the industry are always looking to invent or implement new eco-friendly designs that improve the industry footprint.

Electricity is not always produced in an eco-friendly way, so one of the easiest ways to reduce our industry footprint is to use a solar-powered nursery, which replaces the electrical pump with a solar-powered pump. Most of these solar nursery systems are land-based; however, some water-based solar- and wind-powered systems have been tested.

Another eco-friendly option is the tidal-powered upweller raft (I'm extremely biased because we use it on our farm). No electrical pump is used to accelerate water flow. Instead, the upwelling raft is tied to a mooring in a high tidal flow location. A mouth or scoop is built into the bottom of the raft. This scoop collects tidal flow, which runs into the back of the scoop's wall. With nowhere to go except up, the tidal flow upwells through containers holding the baby oysters and out the top of the raft. On days when the tidal flow is strong, the upweller looks like a whale spouting water.

This type of tidal upweller was invented by a farmer on Martha's Vineyard (Jack Blake—our mentor) and all of the farms on the Vineyard use this method of nursery system to raise our oysters. This design was created out of necessity. Martha's Vineyard property is too expensive for a farmer to own waterfront for a land-based system, and dock space for a floating nursery system is also not accessible. Necessity breeds creativity.

QUALITY OR QUANTITY

ONCE THE FARMER HAS SELECTED THE TYPE of oyster species to farm, they then must decide which type of farmer to be. There are generally two types of oyster farms: those that produce high-quality oysters destined for the raw half shell market, and those that produce lower-quality oysters meant for the shucked meat market. Water quality and regionality of the farm will ultimately dictate whether the product can be of the quality type. To produce quality oysters, first and foremost the oysters must come from clean, desirable waters.

The raw half shell market oysters fetch the highest price but also require a large amount of labor to produce, while the shucked meat oysters fetch a lower price but require the least amount of labor. In the end, both types of farmer make about the same salary. It's a personal choice but also one largely dictated by the location of the farm.

Not all water quality is the same. Some water quality is so poor that it is illegal for oysters to be grown or harvested from a particular area. Some locations have poor water quality yet allow the farming of oysters; however, it is forbidden for those oysters to be eaten directly off the farm and they must first be "cleansed" in clean water before consumption.

Some waters are prone to warm water temperatures, which can breed bacteria or disease and require the oyster to be cooked before consumption. Some locations produce oysters with little to no flavor at all, which are perfect for filling pies, casseroles, or soups.

The best water quality locations have the cleanest of water that highlights the oyster species' best characteristics so that they can be enjoyed raw on the half shell. Eating a raw oyster from a quality farm is a celebration of location and flavor equal to the best glass of wine. Often oysters and wine are paired together for this reason.

The culture method that the farm uses will greatly impact the look and shape of the oyster. The oyster reacts to the culture method throughout its life, augmenting the shell in various responses. The practices that a farmer undertakes will also dictate the appearance of the oyster. This is where the craft and artistry of oyster farming can take on a new level.

TYPES OF
FARM METHODS

ONCE THE BABIES ARE ABOUT ONE INCH IN size, they are ready to graduate from their nursery system to the larger grow-out site.

Generally, there are four types of shellfish farming systems:

- The **bottom culture** system—oysters are placed directly on the bottom sediment with no containers.
- The **off-bottom culture** system—oysters are placed in a cage or container that elevates them off the bottom sediment.
- The **suspended culture** system—gear hangs from a long line or platform down into the water column between the surface and the bottom.
- The **top culture** system—gear floats on the surface of the water.

All four systems can produce amazing oysters. However, the method of choice will ultimately decide the physical characteristics.

This is where the farmer's hand begins to come into play. The taste of the oyster is called merroir, and is strictly dictated by the environment (more on merroir later). The physical appearance of the oyster is completely at the will of the farmer, from genetic traits bred into the shell designs and colors by the hatcheries, to the size and shape of the shells at harvest.

The **bottom culture** method is the oldest and most traditional method of oyster and shellfish farming, as it most closely imitates nature. In the bottom culture planting method, the farmer literally throws the baby oysters directly onto the ocean bottom. The oysters eat and grow while the tide is in, and on tidal farms, the farmer can easily walk onto the flats and harvest the product when the tide is out. On subtidal farms (submerged even at low tide), the oysters are simply dredged off the bottom like the good ol' wild oyster fishing days. This method produces the least uniform-looking oysters with a wide variety of shapes and sizes to their shells, as each animal takes on a different shape while adjusting to the position it has landed in on the bottom. Bottom culture oysters are typically the least expensive, but can also require the least amount of work to produce.

Off-bottom culture is a slight evolution in which the oysters are stored in mesh bags, trays, or bins that elevate them above the ground. On tidal farms, the rack-and-bag method is most typically deployed, as bags filled with oysters are staked from an inch to a foot above the bottom sediment. On tidal farms with large volume, shellfish are hung from platforms or docks, exposed at low tide and feeding during high tide. Either off-bottom method prevents product loss from predation and reduces the amount of sediment accumulation inside the product.

On subtidal farms, off-bottom culture has evolved into the cage system, in which mesh bags full of product are placed inside cages that sit on the ocean bottom until craned up to the surface by boat. Cages can be up to six feet high and six feet wide, utilizing more vertical ocean than the traditional rack-and-bag bottom method. As the off-bottom cage industry continues to grow, there is no reason why these bottom cages cannot become the size of cars or small houses.

The **suspended culture** method is limited to subtidal deepwater sites and has become the preferred way to grow mussels. The shellfish are hung from a support near the surface and grow in the mid-water area of the lease site. Some farms utilize rafts with hanging containers beneath, while others utilize just a horizontal line to support their hanging shellfish gear.

Top culture methods are the newest in the industry and consist of a container filled with oysters that floats on the surface. The oysters live inside the container and are limited to only the food in the surface waters. This method creates the cleanest of oysters, devoid of sediment and shell blemishes; however, this method has the potential for bird waste or other floating hazards to enter the oysters. The floating containers also restrict the ability for boat travel or water use by others and are therefore looked upon as "unneighborly."

The size of the oyster shell is largely dictated by age and the speed of growth. In colder New England waters, it can take three to five years for an oyster to reach three inches in size. In warmer Gulf of Mexico waters, it can take as little as nine months for an oyster to reach three inches. Oysters grow faster in warm water where phytoplankton thrives. In the northern regions, the ocean blooms with phytoplankton in the spring, and like the leaves on a tree, the algae disappear in the colder winter months, forcing the oyster to hibernate until spring when the algae blossom again.

The growth of the oyster begins at the hinge, or point of the shell, and fans outward. The oyster creates the shell in layers, first by shooting out fingerlike appendages and filling the gaps with a very thin calcium carbonate paper-like layer of shell. Over time, additional layers of shell are plastered onto the shell wall from the inside, eventually creating a multilayer hard protective shell. Once the layer is formed, the oyster will again send out the appendages and repeat the process. A thirty-year-old oyster (about as old as they can get) can have a shell over a foot long.

If the layers of shell are chipped while still growing, it teaches the oyster that the environment is very turbulent, and the oyster will continue to plaster layer upon layer to protect itself. This action of chipping the shell ultimately leads the oyster to form a deeper cup. The farmer can use this response to their advantage when shaping the oyster! An oyster with a strong, deep cup is more desirable. The labor of physically chipping the oyster directly impacts the value of the product. A shell that is never chipped will grow very long and thin. A thin, flat shell is considered a less than desirable attribute, as it tends to lead to shell breakage when opening.

Top culture oysters grown in floating bags are constantly chipped by wave action and will form very smooth, small, deep-cupped oysters. Bottom-planted oysters that receive very few shell chips will form very flat, long oysters. Oysters that are run through chipping machines or handled frequently will be somewhere in between. The shape of the oyster is entirely at the farmer's discretion, depending on which techniques they wish to deploy. It is very much an artisan industry.

Oyster aquaculture can also be categorized into two further types of farming: intensive or extensive.

Intensive aquaculture describes the farms who utilize gear culture, such as bags or nets to contain their oysters. Baby oysters are typically purchased from a hatchery as "individual" spat and placed into these containers for grow-out until market size. Intensive oyster farming typically produces your premium-quality half shell oyster.

Extensive aquaculture farming does not utilize containers to grow the oysters but instead most closely mirrors the wild oyster fisheries of the past. The oyster seed is traditionally collected in the wild, on shells or spat collectors, and then cast onto the bottom of the lease site, where it will grow for an extended period of time. Sometimes hatchery seed is used. A dredge is typically used to gather the product off the bottom once the oysters reach market size. On tidal farms of this kind, harvesting consists of handpicking the oysters off the flats.

WHAT MAKES A QUALITY OYSTER?

REGARDLESS OF THE FARMING METHOD, THE taste (merroir) of the oyster is dictated by the environment. When we talk about "quality" oysters, we are referring to the entire range of characteristics bestowed upon that oyster, not only the taste. Non-quality oysters are typically grown in the bottom sediment, dredged up with a giant drag and sold directly to a processor, who opens the oysters and sells just the meats. These oysters are typically canned and used as cooking product for casseroles, po'boys, or soups. They are fabulous oysters, but considered "non-quality" because they do not have unique attributes that make them special enough to enjoy on the half shell. Like a generic table wine, these oysters tend to be used as ingredients and do not serve as the star of the show.

Quality oysters are sold alive, still in the shell. When focusing on a quality oyster, it is the job of the farmer to discover the best attributes their farm site has to offer. Every farm site is different, and this is the beauty of farming oysters. It's through these differences that we get to explore the ocean flavors, discover the nuances of seaweeds and salinities, and begin to understand *what* makes the flavors change.

Discovering the best attributes of an ocean farm is not easy. In fact, the concept of ocean farm attributes is still so new and unexplored that there has yet to be a serious conversation about how a farmer can spotlight the nuances that manipulate their oysters' taste.

When I started farming, merroir and the oyster flavor was the last thought on my mind. I approached the farm like anyone else—as a business. My focus was on margins, potential clients and markets, my business plan, and distribution.

I only began to think about our oyster flavor once the farm was up and running and seemed to be on steady enough ground. As our farm began to produce market oysters, the reaction from consumers became overwhelming. The oyster became an instant hit. We knew we were going to produce a quality oyster because we were lucky enough to farm an environment with some of the cleanest and best growing conditions on Earth; however, we had no idea the site would become one of the best farm areas in New England.

The open waters of the Vineyard Sound are clean, cold, and swift moving. The glacial deposit minerals and sands from the last ice age that formed the cape and islands are unique, non-replicable, and notorious for producing some of the best seafood on Earth. NASA animations of the Atlantic Ocean algae blooms taught us that the majority of nutrients in the Cape Cod and Vineyard Sound waters come down from Greenland and the Gulf of Maine and are met with the Gulf Stream current as it upwells warm-water nutrients from the southern Atlantic. This convergence of ocean currents and nutrients is the reason the Cape Cod region is one of the most productive fishing grounds in the world. It's unique.

When starting to try to decipher the flavor profile of our oyster, we knew the merroir of the Cottage City oyster was on-level with the best, but we also knew our opinions didn't mean much, and frankly, we knew our palates and taste buds were not on par with the professionals. So, we sought out the professionals for help.

Due to the influx of worldwide tourists, Martha's Vineyard hosts some of the best culinary chefs in the industry. Imported to the Vineyard from around the world, these chefs are classically trained and expected to produce high-quality food for high-demand clients. Every professional chef will tell you that great food starts with quality local ingredients. For this reason, we decided very early on to invest in relationships with high-quality chefs.

It's not easy to get a chef to leave the kitchen. By the time summer hits the Vineyard, the work demand is too high to focus on anything else. If you've never hung out with a chef, let me tell you, they are a breed unto themselves. On the Vineyard, we are surrounded by extremely talented chefs. Many come from Michelin-starred restaurants from around the world. Many have worked as private chefs for celebrities and the world elite. All of them share a passion for high-quality food direct from the farmer.

Every chef we have invited to the farm has literally made themselves available on their next day off. Every chef we have brought to the farm has shown a passion for the environment and the life of the farmer. We have come to recognize the chef as the greatest artist in the kitchen, using the farmer's "ingredients" to paint a masterpiece for the consumer to enjoy.

Flavors, textures, colors, and presentation are all tools at the chef's disposal. The type of plate used, the layout of food on the plate, the colors, the tastes, the seasons, and even the drink to wash it down are all thoughts that great chefs think about when designing a meal.

It was one of the best ideas we have had to date. We brought chefs to the farm, showed them the operation, offered them oysters directly out of the water, and then wrote down every word they said. Cottage City Oysters are now one of the most expensive and sought-after oysters on the East Coast of America, and we owe a lot of that success to the chefs who have helped to elevate our product to their clients.

Upon tasting our product, the chefs first describe the merroir as briny. The open-ocean salt is prevalent. The oyster is described as being "high energy," with the first chew releasing an explosion of nuanced sweetness and metallic flavors that quickly end with a clean, bright finish. Hundreds of chefs have told us that the Cottage City oyster is a hallmark New England oyster with a Wellfleet-like brine mixed with the soft minerality and sweetness of the Vineyard landscape, and although this is a huge compliment, it is the environment that gets to take all the credit. Throughout the season our oyster merroir slightly changes in response to the open-ocean phytoplankton that is present. This past winter we discovered the famous blue-green gills within some of our oysters, a sign of one of the most rare and celebrated algae species in existence on Earth. The chef tour always ends with an order to stock their restaurant for the entire season, and leaves us with the job of trying to understand *why* the flavor is the way it is and how to adjust our process as desired.

The input of hundreds of chefs has helped us narrow down a consensus of what the highest-quality oyster we can produce from our farm should be. We have made it our mission to try to design our farm techniques to produce the following attributes in every oyster sold:

The oyster should be about three inches in length. While size can and should vary with every farm, depending on the power of flavor the oyster packs, the unanimous consensus for our product was in the three-inch (three-year-old) range. At this size and age, the oyster is coming into sexual maturity, so the meat is at the pinnacle for flavor, the meat-to-shell ratio is perfect-looking, and the oyster is not yet old enough to be badly blemished by time. Most importantly, our market demands a three-inch cocktail-sized oyster. Being farmed and sold exclusively on Martha's Vineyard, our product has a spot on every party guest list available come summer. From festivals to weddings to backyard gatherings and fine-dining establishments, the Cottage City oyster is typically consumed raw, on the half shell, by best-dressed guests enjoying themselves at an event. If the oyster were too big, covered in barnacles, or dirty in any way, our market would disapprove. Every farm will be different when creating the "perfect" oyster; however, it is typically the market that reveals the answer.

All high-quality oysters should have strong, healthy shells. A thin-shelled oyster will break when shucked, which causes product loss. An oyster with shell damage will also crumble when being shucked. Shell damage is caused by a range of environmental factors such as boring sponges, barnacles, mud blisters, and sea worms, all of which can be controlled. A quality oyster should never break apart when opened. If a bag of one hundred oysters is sold to a chef, all one hundred oysters should be viable product. Product loss is a sign of a lower-quality oyster. Within our operation, we control the cleanliness and quality of shell by air-drying the product.

One of the unique qualities of Martha's Vineyard is that our farms are subtidal. This attribute allows our oysters to feed twenty-four hours a day, but also creates an environment most like the historic natural deepwater reefs: riddled with predators such as oyster drills, worms, barnacles, starfish, and sponges. Unlike their tidal farm counterparts, subtidal farms do not have a low tide that exposes these predators to air, which ultimately kills the predators

and protects the oysters. To counter these issues, we create artificial
low tides by hauling every cage and oyster out of the water, placing
them onto our work rafts, and allowing them to air-dry for twenty-
four to forty-eight hours.

The process of air-drying is a highly orchestrated procedure.
Before the gear undergoes the air-dry, all market oysters from
inside the cage must first be harvested, for after the air-dry period is
complete, the oysters will need at least two weeks to cleanse themselves and reestablish a normal filtering pattern. Once harvested,
the market oysters are promptly iced and head to land for the market. The oysters that are not yet ready for market are air-dried, or as
we like to say, "cleaned by the sun and wind."

The air-drying process is critical to maintaining the highest-quality product. Exposure to air for twenty-four to forty-eight
hours ensures that any harmful biofouling attached to the shell,
such as sponges, worms, or tunicates, cannot take hold. Removing
these objects is crucial to maintaining a blemish-free shell and a
pearl-white interior. Other predators such as drills, crabs, or starfish will either dry out or seek refuge back in the water during
the air-drying process, reducing oyster mortality. The air-dry also
helps to "train" the oyster to hold a longer breath and builds up the
adductor muscle for a sweeter taste. After the air-dry is complete,
the cage is dropped into a new section of the farm and the process
is repeated with a new batch of oysters and cages. We air-dry every
oyster and cage at least four times a year on a rotating schedule to
ensure all market oysters are harvested at peek shell-to-meat ratio.

The outer edge of the oyster shell should be chipped off before
the oyster reaches market. This will ensure that the very thin edge
of new shell growth will not get into the meat when shucked. The
oyster shell should also have a deep cup. The depth of the oyster
shell allows the meat and liquor (filtered ocean water) to sit neatly

within the shell, preventing spillage onto the plate. A flat oyster typically spills the liquor, which can ruin the presentation of the plate. A deep-cupped oyster will also present depth to the plate, an attribute desired by most chefs.

Chefs also expressed interest that the oyster should be power-washed by the farmer. In a high-demand kitchen environment, there is no time to ask staff to clean the oysters. They should be cleaned by the farmer to remove dirt, seaweed, and grime. This is especially true when the oyster is served directly from the farm to the public at a private event or festival. If the farmer is not cleaning the oyster, no one will. The highest-quality oysters are delivered already chipped and cleaned to the end client, with no extra attention needed.

Finally, chefs wanted to see a teardrop shape on every oyster. The teardrop shape is symbolic of the ocean and creates a fine symmetry to the product. When presented on the plate, the similarity in shape of multiple oysters will not detract from the artistry of the chef's presentation.

Strong, clean, teardrop-shaped oysters with a deep cup.

Now that we had consensus on what a quality oyster needed to be, we realized we had to design the farm in a way that would produce these attributes in every oyster, over and over and over again.

We started at the beginning. For our nursery system we choose an ocean-based platform. By utilizing floating bins and a tidal upweller, we could use the natural wave attenuation to bump the oysters into each other, causing micro-chips along the outer edges to help build a deep cup in the first inch of growth.

Once oysters graduate out of the nursery system, we use bottom cages for the next two years of growth to elevate the product above the sifting open-ocean sands to keep them clean and silt-free.

We air-dry the oysters meticulously throughout the season to keep them sponge- and blemish-free. This process trains the oyster to hold its shell closed for long periods of time, extending shelf life.

At the end, we harvest the oysters upon demand and hand-chip every oyster at harvest to ensure uniform size and shape. Hand-chipped Cottage City Oysters are then power-washed to a perfect cleanliness, bagged into two-hundred-count bags, and promptly refrigerated until served to the end client (typically that day).

At a minimum we are neurotic. Like the Japanese *shokunin*, we approach the craft as an artisan or craftsman who dedicates their life toward the pursuit of perfecting the art form. We continue to augment and perfect the system as the site evolves and we learn more about the craft. As crazy as these practices may sound, it has led us to be one of the most talked-about, sought-after, and expensive oysters on the East Coast.

Cottage City oyster with homemade mignonette sauce

Shucking a Cottage City oyster at the raw bar

The open-ocean Cottage City oyster farm off the coast of Oak Bluffs, Martha's Vineyard, Massachusetts

Cottage City Oysters bottom cage hoisted to the surface for harvest and an air-dry

OYSTER GRADES

In most parts of the world, the oysters you receive from the farmer are a smorgasbord of sizes, with only minimum size limits set. A bag of one hundred oysters could contain an oyster as small as two inches, or as large as twelve inches! The harvest size is really at the discretion of the farmer and the demands of the chef.

Menus around the world grade oysters in various ways. An American consumer most likely won't see any verbiage about the oyster size, and if they're "lucky" they might see words like *petite*, *select*, *market*, *XL*, or *jumbo* on the menu, which convey no real meaning to the actual size of the oysters. Japan follows a similar model using words like *high grade*, *top grade*, or *sashimi grade*, which at least conveys some idea of quality.

The U.K., Ireland, and the Netherlands grade their Pacific oysters by weight as follows:

C. gigas
  No. 1 Extra: more than 150 grams
  No. 1: 130–150 grams
  No. 2: 100–130 grams
  No. 3: 85–100 grams
  No. 4: less than 85 grams

China grades their oysters by size, with Bistro (5–7 centimeters), Plate (6–9 centimeters), and Standard (9–11 centimeters) grades.

The French have developed the most intricate and specific method of grading oysters in the world, using a shell-to-meat ratio and, more importantly, a weight grading system. On the menu, oysters are graded from the smallest (size 5) to the largest (size 0000), according to their unopened weight. As the global industry slowly adapts to the French system, you'll begin to see these grades popping up on menus around the world.

Grades vary per species, with just two species currently included:

European (O. edulis) Grades:
    No. 0000: more than 126 grams
    No. 000: 106–125 grams
    No. 00: 96–105 grams
    No. 0: 86–95 grams
    No. 1: 76–85 grams
    No. 2: 66–75 grams
    No. 3: 56–65 grams
    No. 4: 45–55 grams

Pacific (C. gigas) Grades:
    No. 0: more than 150 grams
    No. 1: 111–149 grams
    No. 2: 86–110 grams
    No. 3: 66–85 grams
    No. 4: 46–65 grams
    No. 5: 30–45 grams

MERROIR

WHILE THE FARMER CAN AUGMENT THE appearance of the oyster, the taste of the oyster is entirely dictated by the environment that the oyster is grown in. Place makes taste.

An adult oyster can filter roughly fifty gallons of water a day. They are bivalves, which simply means "two shells." Between the two shells are gills, capable of filtering particles one to ten microns in diameter, about the size of a grain of pollen. As discussed earlier, oysters use their filtering power to trap and consume phytoplankton. This is how oysters accumulate their rich supply of vitamins and minerals for us to consume. Oysters simply extract the nutrients from the already enriched ocean waters and convert it to protein and nutrients our human bodies need.

It's not only the algae that they filter, but also the DNA, nutrients, amino acids, and elements of every substance in their local ecosystem. In this way, the oyster completely reflects the environment it is grown in. The oyster is a time capsule of a region, a process, and an environment. The ability of the oyster to reflect its locality has made the oyster industry most closely resemble the wine industry. Both products are celebrated because of their ability to contain attributes of the location and the process used to create them. In celebrating these differences within the commonality of a product, both wine and oysters have cemented themselves at the culinary table of excellence.

Before we can fully appreciate merroir and the current state of describing taste within the oyster products, we would be wise to first study wine and terroir, the closely related foundation of the concept.

The origin of terroir is thought to have ancient Greek roots. Winemakers of the ancient world began to understand the concept that different regions have the potential to produce very different and distinct wines, even when using the same grapes. The ancient Greeks used this concept to label their wine bottles by stamping them with the region in which the wine was produced, and soon different regions established reputations based on the quality of their wines. This became the origin of classifying wines by the region in which the grapes are grown.

The French built upon this understanding of the ancient Greeks and further evolved the concept by observing the differences in wines from different regions, vineyards, and even areas of the same vineyard. For centuries, much of the wine-making work was conducted by Benedictine and Cistercian monks in the Burgundy region. Vast landholdings allowed the monks to conduct large-scale observation of the influences that various parcels of land imparted on the wine they produced. The monks compiled these works and established the boundaries of these unique areas, many of which exist today as the Grand Cru vineyards of Burgundy.

Through their observations, the French were able to decipher the unique aspects of a place that influence and shape the wine made from it. They created a word for this concept: terroir (from the root *terra* or *territorium* for land). The taste of the land.

Terroir is used to describe the environmental factors that affect a crop's observable characteristics, including unique environmental contexts, farming practices, and a crop's specific growth habitat. When someone refers to the "character" of an agricultural product, they are referring to the terroir.

Wine is the obvious poster child for terroir; however, many artisanal products impart terroir as a means to differentiate. The more popular products for which terroir is studied include wine, meat, cheese, lavender, lentils, honey, spirits, butter, cider, coffee, chocolate, tobacco, vinegar, chili peppers, hops, agave (for making mezcal and tequila), tomatoes, heritage wheat, maple syrup, tea, and cannabis.

Terroir is the basis of the French *appellation d'origine contrôlée* (AOC, or "controlled designation of origin") system created in 1935, which is a certification of authenticity granted to certain geographical indications for wines, cheeses, butters, and other agricultural products, under the auspices of the Institut national de l'origine et de la qualité (INAO), based upon the terroir and a form of geographic protectionism. This is basically a fancy way of certifying where the product was produced according to specific regions or "appellations." The rest of the world follows the French model of labeling such products.

When defining terroir, all wine experts give particular consideration to the natural elements that are beyond the control of humans—the environmental factors.

The main components of terroir include: climate, soil type, geology processes, and the other organisms growing in, on, and around the product.

Climate is typically broken down first into a macro view of the larger global regions and then into a micro view of a smaller subsection of that region. Some vineyards contain smaller microclimates within a single property, and some even observe different climates within a single row of grapes.

Soil types are broken down into the composition and intrinsic nature of the soils, such as fertility, drainage, and heat retention.

Geology processes factor in geological features such as mountains, valleys, and bodies of water, which affect how the climate interacts with the region, and includes the elements of aspect and elevation of the location.

Lastly, terroir looks to the other organisms growing in, on, and around the products, specifically the fauna, flora, and microflora present in the environment, which may act upon or influence the product in some manner. Even the microbial populations living in soils have been described as quantifiable aspects of the overall terroir.

Terroir can be expanded to include the human-controlled elements of the process. Many human decisions in the process of winemaking can lessen or increase the expression of terroir, such as the type of grape grown, when the vines are pruned, how often the grapes are irrigated, the selected time of the grape harvest, the decision to use wild yeasts or laboratory yeasts in fermentation, the type of barrel the wine is aged in, or whether the farmer grew with pesticides or a more organic method.

Terroir stretches to many forms of agricultural production, and the importance of terroir affects the price of the original agricultural product as well as the products created from it.

To describe the environmental influences that can affect ocean-based agriculture, a new word has been created: merroir (from the root *mer* for sea). The taste of the sea.

The concept of merroir is only now beginning to be seriously explored. As ocean farming around the globe has begun to explode, differences in merroir are being compared, and the need to understand the environmental influences on the product are increasing. Like terroir, the mastering of merroir will undoubtedly create a more valued product.

The origin of merroir is unknown. Certainly, the thought of the ocean environmental influences affecting the taste of seafood must be as old as the concept of terroir. For as long as humans have eaten seafood, they have known not to eat from contaminated waters. Even Sergius Orata was known to import oysters from less desirable locations into his farm, so that the more desirable flavors of his lakes could be imparted upon them.

Eighteenth-century British farmers were known to cultivate oysters in specific salt marshes hosting harmless green algae in an attempt to turn the gills of the oyster bright green, creating a much-prized "green oyster," a product of merroir.

Green oysters are now certified "Red Label" in France. According to the French Ministry of Agriculture, the Red Label certifies that a product has certain characteristics that place it on a superior level to that of a similar product. The green algae, responsible for the green coloring of the oyster gills, is a direct example of merroir influence on an ocean product, which in turn creates a higher value for that product.

While the concept of merroir can easily be related to terroir, the finer details of merroir have yet to be explored. The concept is still so new that only a handful of farms have begun to realize the impacts of merroir on their oysters and only a handful of oyster experts have begun to try to describe the nuances that merroir can influence in the flavor.

As a farmer studying the nuances of merroir on an oyster

production system, I can acknowledge firsthand the challenge that is ahead for the industry. It is very difficult to measure the variables that account for merroir. The ocean is a complex system full of environmental influences constantly in flux, yet these marine regions do contain specific merroir flavors. How and why certain marine environments impart specific flavors is still being explored.

In the last few years, cutting-edge science has finally allowed us the tools necessary to count marine species with any sort of precision. Our farm is partnered with Southern Connecticut State University and the Woods Hole Oceanographic Institution to implement environmental monitoring studies to better understand the nuances of our site. Utilizing environmental probes, we monitor the oxygen, temperature, turbidity, pH, and more. One of the coolest tools we use is called eDNA (environmental DNA). All living organisms shed cells, which contain DNA. These cells degrade in the marine environment at certain rates. eDNA allows the scientists to scoop a glass of seawater, filter the strains of DNA, and actually identify the organisms that are present on site. DNA degradation allows us to know approximately how many, and when, organisms were present on the farm. The eDNA program gives us definitive data into the merroir of our site. We are the only farm in the nation that I know of conducting these kinds of studies. Site-specific environmental studies like these are typically underappreciated and too expensive for the average farmer to undertake; however, I believe these types of monitoring programs will become a standard in quality aquaculture farms of the future. Until more environmental monitoring is utilized, the aquaculture industry remains comparable to the times of the ancient Greeks discovering terroir. We have a long way to go, but without a doubt, merroir will become as important to the industry as terroir has to wine. It is only a matter of time.

To date, I have yet to come across a well-thought-out account of what merroir should consist of. Merroir has yet to be properly defined. At the writing of this, a Wikipedia page does not even yet exist! What are the different environmental influences that can affect ocean products, and how do these influences change the flavor of the product?

If the main components of terroir include climate, soil type, geology processes, and the other organisms growing in, on, and around the produce, then we can classify the main components of merroir to include: climate, water type, geology processes, and the other organisms growing in, on, and around the produce.

Climate can be broken down first into a macro view of the larger global regions or "appellations" and then into a micro view of a smaller subsection of that region. An example would be the entire East Coast of America, with a micro region such as New England. Further microclimates would delineate New England into even smaller regions, such as Maine, Massachusetts, Connecticut, and Rhode Island. Even smaller climates would define smaller regions of Massachusetts, such as the Cape and Islands (Martha's Vineyard and Nantucket), Wellfleet, Cape Cod Bay, and the North Shore. Further breakdown would define Martha's Vineyard regions as Katama Bay, Menemsha Pond, Lagoon Pond, and the surrounding Vineyard Sound waters.

Water types can be broken down into the composition and intrinsic nature of the water, such as nutrient densities, free amino acids and algae types available, the amount of tidal flushing (water exchange) a site receives, and salinity. In addition, we should emphasize water quality and the impacts on that quality caused by nearby runoff, moorings, and pollution.

Geology processes include geological features of the shoreline and surrounding runoff areas, such as mountains, valleys, and forests that may contribute runoff, and ocean bottom sediment types such as sand, eel grass, gravel, mud, or bedrock.

Lastly, merroir should look to the other organisms growing in, on, and around the products, specifically the seaweeds present in the environment, which may act upon or influence the product in some manner. Does the production site host a wide array of marine organisms, such as in an open ocean farm, or is the site in a sterile tank in a hatchery? If many species exist, how many, and how much do they influence the product taste? Do oysters grown in a tank with lobsters exude a hint of lobster? If oysters are aged in an oak barrel, will they exude a hint of oak? We are just now beginning to experiment with these influences—my farm being the first known attempt at growing oysters in aged oak barrels. Results to follow soon!

And, as with terroir, merroir should expand to include the human-controlled elements of the process. Many human decisions in the process of oyster farming can lessen or increase the expression of merroir, such as the type of oyster grown, how often the oysters are handled, the type of production system used to contain the oyster, the selected time and size of the oyster harvest, or whether the farmer grew with pesticides or a more natural organic method.

The thoughts on merroir listed above are the beginnings of the industry diving in. Merroir is rarely discussed these days, as most growers have zero knowledge about terroir or what ocean processes influence the merroir of their oysters. Most oyster farmers have not even heard of merroir yet! The term is used by the best of us, those farmers in search of ways to make their product better, and who wish to understand the environments we grow in. As the industry evolves, merroir will become the most important talking point, as terroir has in wine.

On the global scale today, the industry is only now beginning to classify the best oyster regions around the world. Educated consumers are in search of the next Napa Valley of oysters. Like terroir, the root of merroir is grounded in geology. The variations of geologic processes that have formed unique ocean appellations are a defining factor for which regions produce the most sought-after merroir. It makes sense. Unique locations produce unique-tasting oysters.

A level of mass production also matters. Napa Valley wouldn't be known if it didn't produce *a lot* of great wine. Production output matters. A region needs to be able to produce enough great oysters for people to take notice. It takes more than one great farm producing a small batch of oysters. It takes a region of growers to come together around defined themes.

France, Australia, Japan, New England, China, South Korea, and the American West Coast are now recognized as some of the best oyster production regions in the world. Much like the wine appellations around the world, each market is entirely unique and successful. The different merroirs, farm styles, and best practices of these regions are becoming of great interest as developing nations look to stand on the shoulders of these giants as they develop their own aquaculture industries.

Notice that when I discuss the appellations in further detail, I highlight a few of the growers in each region, but the focus is more on the production, merroir influence, and industry practices. These are the leaders, the trendsetters, and the institutional pillars of their respective industries. The areas that produce big numbers have unique products and have the lasting power to see it through to the end.

Like the first world map, the following outline of a global oyster appellation index is drawn for you; however, many of the finer details will need to be filled in. In each of these regions, oyster culture is alive and strong, yet vastly different in nature. The common theme is that the oyster is celebrated for its diversity, nutrition, and tastiness.

I've taken great care to explain the production methods, geographic histories, farm types, and cultural uses, yet the flavors and nuances in merroir are for you to discover. Only you can decide for yourself which oyster is your favorite.

FRANCE

THE FRENCH ARE obsessed with the merroir of oysters and celebrate them as the food of choice for Christmas and New Year's. About half of all oysters produced in France will be consumed between these two holidays. France is the world's fifth-largest producer of oysters with around eighty-three thousand tonnes a year, and the number one consumer and exporter of oysters in Europe, providing 83 percent of all EU production in 2020, mostly through aquaculture.

France is home to the native flat Belon oyster (*Ostrea edulis*). Although the Belon oyster was shipped as far as Rome in antiquity, it was never selected for cultivation. More expensive and harder to find, the Belon oyster only makes up about 2 percent of French production. The majority of French oysters grown today are of the Japanese Pacific variety (*Crassostrea gigas*) and are grown in the traditional tidal flat bottom-planting method, either directly on the sediment or in a rack-and-bag system. Some newer, more progressive farms are beginning to trial hanging nets, floating bags, and bottom-cage systems.

Most oyster farms in France are small and owned by individuals, producing small harvests direct to consumer. French waters are closely monitored and classified into four levels of quality (A, B, C, and D), with A being the best and only waters allowed for the harvest of shellfish. Spat collection is allowed in lesser-quality zones; however, the seed must be removed and planted into A zones, where it is matured for several months before harvest is allowed. The majority of oyster seed in France is acquired through natural wild spat collection.

The French have devised a very clever method of grading oysters, which the rest of Europe has adopted. While the U.S. and Australia use the length of the shell to determine the oyster grade, France uses the weight of the oyster for definition. Grades range from 0 to 5 for cupped oysters (with 5 being the smallest and needing to return to the waters for further growth) or 000 to 6 for flat Belon oysters. The French also grade their oysters using a meat-to-shell ratio. If the meat content is 8 percent or under, the oyster is considered "regular," 9 to 10.5 percent is considered "fin," and over 10.5 percent is "special." Some oysters have been known to have meat content over 20 percent.

Oyster aquaculture in France may end with a production method called "affinage," or finishing, in which oysters are temporarily immersed in marshland ponds called *claires*. This finishing method changes the merroir of the oyster and adds significant value to the end product, especially when the *claire* is considered special or unique.

French oysters are sold au naturel in supermarkets, restaurants, street markets, and huiteries (oyster bars). Ice is used to keep the product fresh until eaten, usually on the half shell with a dab of vinegar mignonette.

Most French oysters are produced from the Atlantic waters, with the Thau Lagoon region being the Mediterranean exception. Favorable merroir in France is broken into six distinct growing appellations: Normandy, Brittany, Pays de la Loire, Charente-Maritime, Arcachon, and the Thau Lagoon. These regions are represented by regional shellfish culture committees or private law structures, like chambers of agriculture or commerce, created in the 2010 modernization of agriculture and fishing law.

Normandy is known for producing fleshy, highly iodized oysters. The most popular oyster varieties from this region are: Utah Beach oysters, known for their sweetness; the oysters of Isigny, known for their pulpy and favorable taste; and those of Saint-Vaast, known for their nutty flavor.

Most sources break the Brittany region into two distinct growing regions, the North and the South. Brittany produces the majority of the farmed Belon oysters in France, which are native to three estuaries on the south coast of Finistère. Other important regions include Cancale, which is considered the oyster capital of Brittany. Cancale produces the Pacific oyster on the tidal flats utilizing rack-and-bag methods. Every other week, the bags are exposed at low tide, when tractors are seen zigzagging between the racks collecting bags of market-ready oysters to take back to the workshops for storage and packaging. Belon oysters are grown farther out to sea directly on the seabed. About forty companies harvest five thousand tonnes of Pacific oysters and one thousand tonnes of Belon oysters annually out of the Cancale region, which has now reached max capacity as all suitable areas have been claimed. Saint-Brieuc, the Bay of Brest, and Morlaix are known for both Pacific and Belon oysters. The Paimpol deep-sea oyster is another favorite.

The Pays de la Loire appellation offers highly regarded oysters from the Île de Ré, Noirmoutier, and the Vendée coast. Grown in rack-and-bag systems on the mudflats, Île de Ré oysters are known for their well-iodized and not too fleshy three- to four-year-old Pacific oysters with defined, deep cups and thick shells. Oysters are graded, power-washed, and then stored for a minimum of twenty-four hours in purification basins of clear ocean water to remove impurities before being sold to markets. Oysters are graded by size, with 4 being the smallest "cocktail" size and 1 being the largest and typically used for cooking.

Charente-Maritime is probably the most famous region of France and is the leading oyster farming basin, accounting for 50 percent of all French oyster production. Around six thousand hectares of oyster farms and three thousand hectares of *claires* call this region home. The slogan "Matured in Marennes-Oléron and nowhere else" is known worldwide. Famed for their blueish-green gills the oysters acquire after ripening in the *claires*, Marennes-Oléron oysters come in four different varieties: the *fine de claire*, the *spéciale claire*, the *fine de claire verte*, and the *pousse en claire*. The *fine de claire verte* and the *pousse en claire* are the only French oysters that have a Red Label.

The most famous French oyster brand, and perhaps the most famous oyster in the world, is the Maison Gillardeau oyster, which calls this region home. Founded by Henri Gillardeau in 1898, the brand quickly became a favorite among the French and for four generations has been deemed the most prestigious oyster from the Marennes-Oléron pond in Charente-Maritime. The brand has become so famous that Chinese farms began to sell counterfeits, so Gillardeau oyster shells are now laser engraved with the company logo to ensure authenticity.

The ancient Arcachon beds that played such a strong role in French history continue to produce to this day and remain one of the strongest breeding centers for most of France's oyster basins. Today, the bay has about twenty-six oyster farms, with the top four oyster varieties being Le Banc d'Arguin, Le Cap Ferret, Le Grand Banc, and L'Île aux Oiseaux.

The outlier of the bunch is the Thau Lagoon appellation in the southeast Montpellier region of France. The Canal du Rhône à Sète links the lagoon to the Rhône river, while the Canal du Midi connects the lagoon to the Atlantic Ocean. The Thau Lagoon also has access to the Mediterranean Sea and is fed by France's largest hot spring, the Balaruc-les-Bains, which contains large amounts of trace elements and essential mineral salts. The lagoon is classified as grade A waters and is France's second-largest lake at twenty-one kilometers long and eight kilometers wide. The entire lagoon is pretty much a series of locally owned oyster farms utilizing "oyster tables," dock-like structures with the oyster bags or strings suspended beneath. Eighteen varieties of shellfish are found in the lagoon with 750 oyster producers using 2,750 oyster tables and producing around 13,000 tonnes of oysters each year (over 8 percent of total French consumption). The flat Huîtres de Bouzigues variety is regarded as the oldest in the region and named after the village where production was started. La Maison Tarbouriech, a "high-tech" grower in the region, invented and utilizes the Marée Solaire, a solar-powered device that raises and lowers the strings of oysters at various time frames to produce an artificial "solar" tide air-dry, creating a new twist on a traditional practice in the region.

AUSTRALIA

MODERN HUMANS (*Homo sapiens*) reached Asia by seventy thousand years ago before moving into Australia; however, *Homo erectus* had already been in Asia for at least 1.5 million years. A third species, the Denisovans, were also known to inhabit the Asian region, and Melanesians and Aboriginal Australians carry about 3 to 5 percent of Denisovan DNA. This is explained by the breeding of modern humans with Denisovans as they migrated through the Eurasian continent towards Australia and Papua New Guinea.

The ocean has always separated Australia from the Asian continent, but sea level has drastically changed overtime, creating periods of opportunity for daring seafaring explorers.

Whatever the motivation, human occupation of Australia is thought to have occurred between fifty thousand and sixty-five thousand years ago, highlighting one of the greatest maritime achievements of early humans. Recently a possible shell midden has been discovered at the site of Moyjil in Warrnambool, Victoria, which would date human occupation in Australia to as early as 120,000 years ago; however, additional data is needed and investigations continue. Regardless of the exact age, the midden provides evidence for the consumption of shellfish by early humans in the region.

Due to the perishable nature of ancient boatbuilding supplies, archaeological evidence of what the ancient sail crafts looked like will probably never be discovered, though most likely the rafts would have been made out of bamboo, the most common boatbuilding material in Asia.

One of the most important details to note is that the Aboriginal Australians on the coastal regions feasted on oysters and cockles to heart's content, as evidenced by the thousands of middens decorating the Australian coastline, some dating back as old as twelve thousand years ago. Two main indigenous oyster species are found within the middens: the Sydney rock oyster, which occurs in the warmer waters, and the Angasi flat oyster, which lives in the colder waters surrounding the continent. The oldest shell middens are most likely located under the ocean today, as sea levels have risen since ancient times. There is evidence of elaborate trade networks that existed thirty thousand years ago, where seashells (used for decoration) were traded as far as five hundred kilometers inland.

In 1606, Dutch navigator Willem Janszoon landed on Australia's northern coast, followed by twenty-nine other Dutch navigators who explored Australia in the seventeenth century, eventually naming the continent New Holland. Other European explorers navigated the coastlines until 1770, when Lieutenant James Cook claimed the land for Great Britain.

In 1788, the first British penal colony was established and

European disease quickly followed. The Aboriginal populations plummeted. So too did the wild oyster stocks. The use of oyster shell in the production of quicklime concrete quickly depleted the oyster population. As an example of the destruction, parliamentary records show that from just five bays in Tasmania, more than twenty-two million Angasi oysters were harvested in the 1860s.

The government quickly stepped in and introduced cultivation methods based off the French techniques in 1872. This proved unsuccessful, due to predators and silting that formed within the ponds. Tasked with the development and success of the farms was Thomas Holt, who was appointed as chairman of a royal commission into oyster culture, which passed a law in 1880 to allow the formation of private oyster leases. Methods of farming oysters on raised stones or in cages that raised the oysters above the seabed eventually proved successful, and the industry has been off and running since.

Australia is only the world's twelfth-largest producer of oysters, with over six thousand tonnes produced in 2018 (prior to Covid). While their production numbers lack luster, it is the unique species of oysters being farmed in Australia that make it one of the most prized regions for oyster merroir. Hundreds of miles of natural rivers, estuaries, and bays make for some of the most unique growing areas in the world; however, these fragile habitats also make it the most vulnerable to climate change. Fires, floods, and disease have wreaked havoc on the industry recently, killing millions of oysters and forcing farms out of business.

The Japanese Pacific oyster was introduced to the continent as early as the 1940s, threatening the survival of the native species. Today, the Pacific oysters are produced in Tasmania and Port Stephens, though the rest of the country views these oysters as a "noxious fish." The native Sydney rock oysters are farmed in New South

Wales, southern Queensland, and Albany in Western Australia, with Wallis Lake and the Hawkesbury River being the main production regions in the country. The Angasi flat oyster is farmed on the south coast of New South Wales and carries the highest value of any oyster produced in the country.

Australian oysters are predominantly eaten raw on the half shell and served year-round, with the Angasi oyster's peak season being from May to November. Famous cooked oyster dishes include the Rockefeller and Kilpatrick recipes, which consist of oysters cooked in their shells, topped with crispy bacon and Worcestershire sauce, usually served as an appetizer.

There are over forty-four growers of oysters in the Tasmania region, the majority of which produce the Pacific oysters. Growing areas include the north, east, and southeast coasts. Tasmanian oysters are famed for their large, elongated shells and tastes that range from briny to cucumber-like.

The majority of production in Queensland comes from the cutting-edge Moreton Bay region in South East Queensland. Sydney rock oysters grown in off-bottom cages can be found off the island of Moreton within the open-ocean environment, as opposed to the traditional rivers and estuaries, creating a truly unique merroir. There are also now small trials of growing the tropical Blacklip rock oysters in the region.

New South Wales creates the most oyster production in the country with over 280 farms in the region. The Wallis Lake region just north of Sydney is world famous for their native Sydney rock oyster varieties. This enormous lagoon covers roughly twenty-five thousand acres when full and is surrounded by protected lands and the Wallingat National Park. Farming in the lake region began in about 1900, and modern farmers use stakes and frames covered in cement to attract wild set oyster larvae. These oysters grow on the stakes for about three years before they are knocked off and then allowed to grow on the estuary bottom for another six months to a year. It takes at least three years to bring a Sydney rock oyster to market. The merroir is celebrated as being full of brininess, creaminess, sweetness, minerality, and umami—Australian oysters at their best.

Port Stephens produced about 16 percent of the state's supply of Sydney rock oysters until 2022, when QX disease wiped out 100 percent of all Sydney rock oysters in the region. Farmers have had to rely on their supply of Pacific oysters, which are also facing large mortality rates due to an unexplained disease.

In the southern regions of New South Wales, the native and truly unique Angasi flat oyster species is making a comeback. Grown within the bottom substrate or using the off-bottom method, this flat oyster takes the merroir of sweet, nutty, gamey undertones. With a slow growth rate (three to four years to maturity) and a short shelf life, the Angasi oyster is quickly becoming an oyster connoisseur's most sought-after Australian treasure.

JAPAN

JAPANESE OYSTER SHELL middens date from 14,000 BC to 1000 BC, a record of how long the bivalve sustained the ancient Japanese people.

Multiple sources state that oyster cultivation began in Japan as early as 2000 BC; however, these are unsubstantiated claims. Documented proof of oyster cultivation in Japan is thought to have begun in the Muromachi period (1336 to 1573 AD), and records state cultivation methods were in use in the western Hiroshima area during the Tenbun era (1532 to 1555 AD). Since the 1500s, Hiroshima has remained the main region for oyster production in Japan.

The most famous story of Japanese cultivation is said to have occurred in 1624, when fisherman Heishiro Yoshiwaya from Hiroshima recognized that baby oysters would attach to rocks and the bamboo of his fish traps. These two observations are believed to be the basis for the ancient Japanese method of using stones (*ishimaki*) and bamboo sticks (*hibitate*) to cultivate oysters.

The ancient *ishimaki* method of cultivation was similar to the method developed by Sergius Orata, in which stones were laid into a muddy bottom area and, out of necessity, the baby oysters settled onto the stones. The oysters were grown to adult size and sold into the marketplace.

The *hibitate* method used the same principles, but the stones were replaced with branches of bamboo. This allowed more surface area for oyster spat collection and allowed the farms to move into deeper waters.

Yoshiwaya experimented with the two cultivation techniques in the Hiroshima area with great success. His *hibitate* method proved most productive and oysters were sold to other provinces via the ancient Sanyo trade route. In the 1670s (Enpō era), transportation by "oyster boats" advanced into Osaka, where water channels developed to deliver Hiroshima oysters.

Osaka's location along the Tōkaidō and Nakasendō trading routes made it a center of manufacturing and distribution and one of Japan's first capitals. Osaka has long been called "Water City" and "*happyaku yabashi*," meaning "808 bridges."

Japan is the world's third-largest producer of oysters, producing around 176,000 tonnes in 2018 (pre-Covid). As one of the oldest oyster-producing regions in the world, Japan offers a unique mix of traditional and modern harvest practices.

The geography of the Japanese islands are unique, formed as the result of several large ocean movements occurring over hundreds of millions of years. Originally attached to the Eurasian mainland, the islands were pulled away from the continent by plate tectonics, which created the Sea of Japan about fifteen million years ago. The deepwater upwelling nutrients, trenches, and currents surrounding the islands produce oysters like no other place on Earth.

In Japan there are four distinct merroir regions: Hokkaido, Miyagi, Hiroshima, and Iwate. There are over a dozen species of oysters along the coast of Japan. Of these, *Crassostrea gigas* (Pacific oyster) is the most important one, with a growth rate higher than that of any other species and a taste considered to be one of the most flavorful. They are considered "in season" from November to April but can be found year-round fresh, frozen, or cooked. Oysters in Japan are eaten in every conceivable way possible—raw on the half shell, sushi style, steamed, fried, grilled, in soups, in rice, and even simmered in oil.

Near-shore oyster farms in Japan utilize the rack-and-bag system, staking bags into the bottom sediment. However, most of Japan utilizes a suspended type of oyster cultivation method due to the great depths of ocean surrounding the region. Instead of growing oysters on the shallow mudflats, the floating and mid-water growing methods are used, in which vertical and horizontal lines attach to floating rafts and suspend some fifteen meters below the surface water. Oysters grow attached to the suspended lines.

The vast majority of oysters are grown for the shucked meat markets, where shell design matters little, but recently demand for oysters sold on the half shell has risen, ushering in a new wave of farms around the country to meet demand. Many of the farms that were devastated in the 2011 tsunami used the reset to incorporate modern rack-and-bag methods, with the help of Australia and France. These methods produce higher-quality-looking oysters over the suspended model, and have become popular varieties for the half-shell market.

The most productive oyster regions of Japan also have their own famous oyster varieties. In northern Japan, the cold sea around Akkeshi Bay in the northernmost part of Hokkaido provides ample food and optimum temperatures, making it possible to harvest Pacific oysters year-round. Served fresh, raw on the half shell, Kakiemon brand oysters are known for their depth of flavor and plump, juicy flesh.

In the northeast Tōhoku region, located on the Sanriku Coast lies Miyagi, famous for the Maruemon oyster, cousin to the Hokkaido Kakiemon oyster. It is the second-largest oyster-producing region in Japan, and eleven fishing villages make up the appellation, where farming dates back to the 1600s. Utilizing a "drooping chain" off-bottom method of farming, lines of oysters are hung from dock-like structures to filter the high tides and sunbathe in the lows. While "in season" (autumn to winter), most oysters from the region are eaten fresh, raw on the half shell. In Iwate, the succulent Hanamigaki oysters are also available in spring and grow three times larger than winter oysters.

Hiroshima is the top producing region of Japan, producing thirty thousand tonnes of oysters annually. Oysters are grown on the Seto Inland Sea. During peak oyster season (starting late January), the Hiroshima Oyster Road opens along the coast with numerous dining establishments dedicated to the local bivalve.

When someone hears of oysters from Hiroshima they can't help but think of the atomic bomb that was dropped on the region on August 6, 1945, to end World War II. This is obviously a marketing nightmare for the producers in the region.

According to the City of Hiroshima website, "The radiation in Hiroshima and Nagasaki today is on par with the extremely low levels of background radiation (natural radioactivity) present anywhere on Earth. It has no effect on human bodies." The website further explains that

*The initial radiation emitted at the moment of detonation inflicted great damage to human bodies. Most of those exposed to direct radiation within a one-kilometer radius died. Residual radiation was emitted later. Roughly 80% of all residual radiation was emitted within 24 hours. Research has indicated that 24 hours after the bombing the quantity of residual radiation a person would receive at the hypocenter would be 1/1000th of the quantity received immediately following the explosion. A week later, it would be 1/1,000,000th. Thus, residual radiation declined rapidly.**

Emboldened by these facts, modern Hiroshima oyster farmers go one step further to ensure their oysters are purged of any contaminants—they purify the oysters over a day within highly controlled tidal pools before their sale. This purge guarantees the removal of contaminants, but also removes the oyster's characteristic astringent taste.

CHINA

LARGE-SCALE CULTURE OF oysters started in China during the 1950s; however, 1985 marked the start of China's rapid growth in the aquaculture realm when the State Council issued the *Directive on Relaxing Policies and Accelerating Aquaculture Development*. These policies brought major investments to the industry by the early 1990s, spawning enormous farm growth across all regions of the coastline and cementing China as the largest global producer of farmed shellfish. In 2018, the National Health Commission of the People's Republic of China added oysters to the first batch of

* "Q. Is There Still Radiation in Hiroshima and Nagasaki?" City of Hiroshima, city.hiroshima.lg.jp/site/english/9809.html.

healthcare products listed as "both food and medicine."* In China, the oyster is revered for its health qualities and nutritional value.

Today, China employs over one hundred thousand long-term workers in the oyster industry and produces over 80 percent of the global production. By far the largest producer of oysters in the world with over 5.4 million tonnes produced annually, China's oyster production is focused solely on quantity. Hundreds of millions of oysters are harvested to market annually as economically as possible, utilizing farms that are tens of thousands of acres in size, state-of-the-art machinery, and a labor pool that amounts to that of a small army.

Statistics on the Chinese industry can be almost impossible to find, as the government tends to keep its cards closely guarded. There are more than twenty species of oysters in China. The three most-cultured species include the Pacific oyster, Portuguese oyster (*Crassostrea angulata*), and Hong Kong oyster (*Crassostrea hongkongensis*). The vast majority of oysters consumed in China are eaten cooked, enjoyed roasted or steamed, but also used as ingredients in soups, porridges, hot pots, salads, and oyster sauce, the staple ingredient in many Asian recipes.

China's coastline extends about eighteen thousand kilometers, crossing the tropical, subtropical, and temperate climate zones, which makes the farming of oysters, clams, scallops, razor clams, mussels, cockles, abalone, and sea snails viable. Pacific oysters are predominately reserved for the northern regions, while the other species are

---

* Chang-shi Zhou, Chang-kun Huan, and Yong-tong Mu, "Development Strategies of Oyster Industry in the Coastal Region of Pearl River Delta: Based on In-Depth Interviews of Mariculture Operators, Enterprises, and Government Administrative Agencies," *Research of Agricultural Modernization* 35 (2014): 757–762.

mainly cultured in the southern regions of China. Shandong is the largest oyster production region in the north, while Fujian, Guangdong, and Guangxi make up the largest southern appellations.

Hatchery technology exists for oyster production in China, but the majority of oyster farms rely on the capture of wild oyster spat. The reliance on natural seed sources can cause major swings in production based on the success or failure of spat collection for any given year and has forced the local governments to enforce strict rules on water pollution in spat collection areas.

The spat is typically collected from the wild onto two- to four-foot lines that contain concrete plugs about every six inches. Once the wild baby oysters are attached to the concrete plugs, the lines are transferred to a suspended grow-out system utilizing either rafts or buoys for growing until they are market size, about three years old and usually over four inches.

The northern Shandong appellation produces the Pacific oyster variety predominantly in the Liaoning province, off the island of Guanglu in the Yellow Sea. Province officials have created the slogan "Increase fishermen's incomes, go green, improve quality."* These goals have set into motion some of the largest and most productive oyster farms in the world.

With rows of buoys extending out of Liutiaogou Bay like fields of wheat, over 360,000 hectares of water have been converted into oyster farms able to be seen from space. Fleets of fishermen navigate their wooden vessels between the floating rows of buoys to harvest the lines of oysters suspended below. A boat crane lifts the

---

* Mark Godfrey, "Home Province of Chinese Processing Port Dalian Sees Jump in Imports, Value," Seafood Source, seafoodsource.com/news/supply-trade/home-province-of-chinese-processing-dalian-sees-jump-in-imports-value.

horizontal grow lines above the deck, where fishermen quickly strip the oysters free. Working down the line, the vessels fill with tons of oysters quickly. Once filled to capacity, the fleet lines up at the Guanglu island dock, where massive cranes unload the cargo nets full of oysters from the overloaded decks. The vessels return to the fields to repeat the harvest process, trying to keep up with the productivity of Mother Nature.

The Fujian region of China sandwiches a narrow strait that connects the East and South China seas between the mainland and Taiwan. Called the Taiwan Strait, this body of water is legally split in half, with the Chinese Fujian region and Taiwan each claiming ownership of their respective sides. Regardless of jurisdiction, the strait contains some of the most productive seafood waters in Asia.

When it comes to seafood harvest totals, Taiwan ranks first in the world for Pacific saury (mackerel), second for tuna, and third for squid. If Taiwan were its own country, it would globally rank fourth in the world for oyster production (between Japan and the U.S.), with about 160,000 tonnes annually. The majority of oyster production in Taiwan comes from the southwest coast within the Taiwan Strait, stretching from Hsiangshan in Hsinchu County in the north, to Tungkan in Pingtung County in the south. Cultivation methods have remained the same since the 1940s.

Growing predominately the Portuguese oyster variety, Taiwanese farms utilize a bottom-growing system in which bamboo poles about three feet high are jammed vertically into the mudflats. Recycled oyster shells are tied onto the poles and used to catch the wild oyster spat. When the tide rushes back in, the poles become submerged and the wild oyster spat chooses the recycled shell as the ideal substrate to settle on. In this way, millions of oyster larvae are collected and grown to market size. Oyster farmers in the region can be seen driving makeshift three-wheel tractors on the mudflats at harvest time. Market oysters are broken off the bamboo lines and driven back to their respective shucking facilities.

In the deeper waters off Taiwan, farmers are deploying raft culture similar to their Chinese counterparts. Taiwanese oysters are grown exclusively for the shucked meat market as none are considered fine enough for the raw or half-shell market.

The Guangdong appellation in China is known for the famous Hong Kong oyster species, which is found naturally around the city and the mouth of the Pearl River. The most famous oyster city in the region is Lau Fau Shan, where the shucked and dried "golden oyster" has been harvested for more than seven hundred years. Traditionally harvested from the muddy banks of Deep Bay, the oysters are shucked and the meats are spread out across wire mesh screens to allow the sun to dry them to perfection. As many as one hundred tonnes of dried golden oysters are harvested annually, down from the almost three hundred tonnes produced in the 1960s.

Since the 1970s, golden oysters have been grown with a suspended farm method, utilizing over eleven thousand bamboo rafts and hanging lines within Deep Bay, a waterbody surrounded by marshland and the city of Shenzhen, which harbors more than thirteen million people today along the shore. Historical evidence teaches us that industrial growth typically means the downfall of water quality, which has become the case in Deep Bay. Shenzhen's government banned oyster farming in their half of the bay in the early 2000s citing pollution as the main culprit.

On the Hong Kong side of Deep Bay is the Mai Po Nature Reserve (the only nondeveloped nature preserve in the region). Neighboring the Nature Reserve is the famed village of Lau Fau Shan, which was able to continue small-scale harvests until 2023 when water quality in the bay finally reached a tipping point. Red tide, a harmful algal bloom, has become commonplace, killing the majority of the remaining oysters. Government and nonprofit pilot projects are being implemented to save the industry and return Deep Bay to cleaner waters. Only time will tell if these endeavors succeed.

An estimated 70 percent of China's oyster production comes from Qinzhou, a city in the southernmost appellation of China's Guangxi Zhuang Autonomous Region. Autonomous regions in China have their own local governments (sort of like a state or province). The Guangxi region borders Vietnam and is known for its maze of rivers, caves, towering mountain formations, and lush natural landscapes.

Qinzhou's oyster farms cover over ten thousand hectares and produce over $400 million annually. The oysters are grown within the Maoweihai natural harbor using a suspended grow-out system in which giant bamboo rafts anchor to the harbor bottom and fill every foot of the meandering waterway. With a raft culture system of this magnitude, boats are unable to navigate the waters with precision, so the industry is less reliant on mechanical cranes and instead utilizes a giant manual labor force. Workers are "delivered" to the rafts, where they spend the day lifting oysters from below, severing them from the grow lines and loading them into vessels for transport back to port, all the while balancing on the thin bamboo shafts like high-wire walkers. A slip through the bamboo raft into the sea below would be a costly error, ending in a myriad of cuts from the razor-sharp oysters beneath.

The popularity of Qinzhou oysters has motivated the city to host an annual oyster festival for the entire month of December, inviting tourists from all over the world to celebrate the oysters and to consume a variety of oyster dishes prepared by chefs on the spot.

Arguably China's most famous oyster contribution is the creation of oyster sauce, a delicious condiment used around the world to flavor Asian cuisine. Invented around the 1870s, traditional oyster sauce was produced by boiling oysters in iron basins for about thirty minutes. The oysters were removed and the leftover watery liquid was further reduced until a thick syrup was created. Salt, water, sugar, and soy were added to the syrup. Today most oyster sauce is a combination of sugar, salt, and cornstarch. Oyster extract and soy sauce are now added for additional flavor.

KOREA

KOREA IS THE second-largest producer of oysters in the world with just over 303,000 tonnes a year in 2018 (pre-Covid). Used in everything from soups, rice dishes, ramen, and porridge, the oyster is a protein staple in the Korean culture.

The unique geography of the region allows for production on the flats, in subtidal waters, and in deep sea. This wide range of farmable habitat allows for the entire gambit of farm systems. The farms are nationally protected from pollution and go through biannual inspections by the U.S. Food and Drug Administration, made mandatory from the 1972 MOU Regarding the Safety and Quality of Fresh and Frozen Molluscan Shellfish between the U.S. and Korean governments. More than 50 percent of the oysters produced in Korea are exported to Japan and the United States. This makes Korea the largest exporter of oysters in the world. The exported oysters are typically shucked meats used for cooking.

The vast majority of oysters produced in Korea come from the seaside cities in South Gyeongsang such as Tongyeong and Geoje. The South Gyeongsang coastline enjoys relatively calm waters throughout the small bays and islands, allowing for perfect oyster farming conditions on a large scale.

Oysters are grown using the suspended longline system, where young oysters hang attached to horizontal longlines that remain submerged in the sea until harvested. After harvest, the oysters are taken to processing plants where they are opened and the meats are thoroughly washed. Oyster meats are traditionally sold in auction by the truckload to fish brokers and then delivered to markets and restaurants.

Popular oyster dishes in Korea include the oyster pancake, oyster seaweed soup, braised oysters, and spicy raw oysters. Each dish can contain up to thirty oysters!

Tongyeong, nicknamed the "Naples of Korea," is located in the middle of South Korea's southern shoreline. Surrounded by hundreds of tiny islands, the oyster farms here are so massive that they can be seen in satellite images from space. Harvested in the winter from November to February, these oysters are grown to be as big and meaty as possible. A majority of the oysters are of the triploid stock to help achieve plump meats. Oyster sizes are large, four inches or more with a subtle flavor, less briny than European oysters, and the meats are gray and cream colored. While the merroir is lesser known in this area due to the majority of oysters harvested being processed and cooked, the more than 250 oyster farms and farmers proclaim that they grow the best oysters in the country.

Geoje produces Pacific oysters between the months of July and December, harvested from vast ocean farms using cranes and barges. This is truly the epitome of industrial oyster farming. Utilizing suspended rope or net systems, the farm teams extract the product from the waters and place the bundles of oysters into large floating conveyor belts, which break the bundles of oysters apart and give them a quick power-wash. The oysters are then packed into large containers and taken back to shore, where they are dumped onto shucking tables surrounded by mostly women shuckers, who are known to shuck faster than men due to better dexterity and precision. The women make quick work of the oysters, opening about one every two to three seconds. The meats are collected in pails and then run through a giant washing and sanitizing system. At the end of the assembly line, meats are inspected for damage or shell before they are packaged in final containers with sterile seawater.

UNITED STATES

THE MOST RECENT and rapid shift in oyster culture has occurred in the U.S. over just the last decade or two. Seemingly overnight, oysters have transformed from a canned and anonymous meat product to a social appetizer with individual brands known for their flavors and locations. Oyster culture in the U.S. has entered a renaissance age, in which the best regions, tastes, varieties, farm methods, and love of oysters are being rediscovered.

Being the fourth-largest producer of oysters globally and containing the second-largest shoreline on Earth (Canada is first), the United States offers a wide range of unique and diverse coastal habitats and one of the fastest-growing aquaculture industries in the world. The U.S. is forecast to pass Japan and become the third-largest oyster producer in the coming decade as demand from younger generations continues to accelerate the American industry.

U.S. consumers have begun to distinguish the variable qualities in oysters, and in return we have seen the industry consolidate into a handful of large production appellations, each with their own markets and merroir, very similar to how the craft beer or wine industry developed. While taste and merroir can be subjective, production numbers are key. To be a "Napa Valley"–type appellation, producers must output enough product that it can reach the far corners of the Earth. It must not only be great tasting, but also be produced on a scale that allows consumers to grasp it.

For a detailed account of the North American producers, I suggest Rowan Jacobsen's book, *A Geography of Oysters: The*

Connoisseur's Guide to Oyster Eating in North America. Jacobsen details the regions and highlights the most famous oyster growers from those regions. Unfortunately, some of the growers highlighted have since closed shop.

Oyster farming is a hard profession with a lot of turnover. It takes dedication, hard work, skill, and a little luck to succeed. The majority of oyster farms in the U.S. are part-time gigs, people with second jobs to help support their farming endeavors. It takes a decade or more to become a successful, established farm, able to support a family. I could make a list of all the farmers producing oysters today, but in just five years more than half would be replaced with new growers and brands. As the industry marches forward, only the most dedicated growers persevere, taking a decade or more to reach large enough production numbers for their brands to become noticed. For those who can't find success, a waiting list of new growers quickly fill the void to stake their own claim.

Instead of looking at the list of ever-changing growers in the U.S., below I have taken a more macro approach to the American appellations by highlighting the regions that, as a farmer, I have distinguished as being unique in flavor profiles, geography, and growing techniques. These are the areas that are producing highly distinct merroir, unique products, and recognizable brands.

New England

NEW ENGLAND HOSTS four very different merroir regions: Massachusetts, Maine, Connecticut, and Rhode Island. Oysters grown on the East Coast of the United States (from Canada to Florida) are all of the *Crassostrea virginica* variety; however, the merroir of each appellation is as unique as can be. Each state is further broken down into important regional distinctions.

Below I highlight the most popular, most productive, and most unique growing locations I have found to date. As a New England farmer, I am undoubtably biased, but I believe this region produces not only the best oysters in America, but some of the best oysters in the world. This fact is due not only to the experience and industry-leading techniques developed in New England, but more importantly to the geology that created the New England region.

Bedrock analysis and fossils tell us that the New England region sits on the edge of the Laurentia Craton, the North American continental crust that has remained relatively stable for the last six hundred million years. This crust is one of the oldest on Earth. Initially, volcanic activity covered the New England area in a bedrock of granite and silica-rich sediments, which are still present today. About 470 million years ago, the North American crust collided with a smaller micro-continent called Avalonia, which contained modern-day Newfoundland, the British Isles, France, Nova Scotia, New England, and northwest Africa. Interestingly, these geographies are all some of the best modern-day oyster appellations, sharing much of the same mineral composition and old bedrock.

As the two continents crushed together, the ancient seabed uplifted thousands of feet to create western Massachusetts, Maine, New Hampshire, Connecticut, Vermont, and the Appalachian mountain range (extending from Canada to Alabama). Today, these ancient seabed materials now erode into the rivers and coastlands, providing modern-day New England oysters with a plethora of ancient minerals found nowhere else on Earth.

About two hundred million years ago, the super-landmass Pangea was created as the Gondwana supercontinent—consisting of Africa, South America, and Asia—collided into the Laurentia Craton. This collision sandwiched Avalonia until one hundred million years later when the continental plates split and Pangaea broke apart. The Ireland and United Kingdom half of Avalonia was ripped off the East Coast of North America in a northern direction while Africa spun off to the east. A sliver of the old Avalonia continent remained attached and is now the landmass under the New England coastline. This is why the New England bedrock is the same as Ireland and the United Kingdom, because they are formed out of the same ancient continent!

If these events weren't unique enough, the last 2.5 million years of ice ages have defined New England further. At times buried beneath a mile of glacial ice, New England was compressed by the Laurentide Ice Sheet's growth and melting cycle until as late as just twenty thousand years ago. These glacial events have scoured the ancient bedrock and coastlines, bringing rich deposits to the coastal zones and even forming New York's iconic Long Island and the Massachusetts shoreline and islands. These landscapes are some of the newest on Earth.

As the glaciers receded, a new, pristine region built on the most ancient bedrock of Earth was revealed. Trees, oysters, and even humans have populated the area in just the last twenty thousand years. The New England region is one of a kind, and the oysters found within are like no other, prized around the world for their salty and minerally tastes.

Massachusetts

THE MOST FAMOUS oyster in New England comes from Wellfleet, Massachusetts, a small harbor within the arm of Cape Cod Bay. The entire arm of Cape Cod was formed about twenty thousand years ago from glacial deposits; however, with shifting sands and rising oceans, the modern coastline would not have been distinguishable until about three thousand years ago. This sand is generally what makes the Massachusetts oyster unique, as the glacial deposit sandy coastline provides a hard bottom for farms unlike any other state.

Discovered as early as 1606, Wellfleet was named "Port Aux Huitres" (Oyster Port) by French explorer Samuel de Champlain. The name was later changed to Billingsgate by the earliest European settlers, who adored the widespread oyster beds and made use of them at once, and continued to do so until their extinction. In an essay written by Levi Whitman, dated 1793, Whitman gives his opinion that

> No part of the world has better oysters than the harbor of Wellfleet. Time was when they were to be found in the greatest plenty, but in 1775 a mortality from an unknown cause carried off the most of them. Since that time Billingsgate oysters have been scarce, and the greatest part that are carried to market are first imported and laid in our harbor, where they obtain the proper relish of Billingsgate.*

Nearly fifty years later, in 1841, Augustus Gould wrote in *Invertebrata of Massachusetts* that

> They say that Wellfleet, where the southern oysters are planted for Boston use, was originally called Billingsgate, on account of the abundance of fish, and especially of oysters, found there; that they continued to be abundant until about the year 1780, when from some cause they all died; and, to this day, immense beds are shown there, of shells of native oys-

---

* Levi Whitman, "A Topographical Description of Wellfleet" in *Collections of the Massachusetts Historical Society*, 3rd ed. (Boston: Massachusetts Historical Society, 1794), 119.

*ters which perished at that time. They say that, before that
time, no such thing was thought of, as bringing oysters from
the south.**

Wellfleet supplied Boston with oysters until the last reef was
harvested in 1775. It was then used as a relay point because of the
hard sand bottom, to which seed and lesser-quality oysters from
other regions were shipped and allowed to filter for months to
acquire the famous Cape Cod salty, minerally, lobster bisque taste.
Today, bottom-planted aquaculture farms have replenished the
Wellfleet harbor, and the Wellfleet oyster stands as a classic New
England merroir.

Just across the Cape Cod Bay lays Duxbury, home of Island
Creek Oysters, the largest oyster company in Massachusetts. An
umbrella organization of Duxbury farms, Island Creek has dominated the marketplace by becoming the largest distributer of Massachusetts oysters. They not only sell their bottom-planted Duxbury
brands but have begun to include oysters from the Massachusetts
islands, Connecticut, and Maine. Island Creek Oysters has their
own oyster campus, an awesome destination for locals and tourists
alike, to see firsthand how the farms and hatchery processes work.
In addition, Island Creek has restaurants throughout the state,
caviar products, canned goods, and stores along the New England
coast. As a fellow Massachusetts grower, I can't say enough good
things about them. We are lucky to have them distributing the
Massachusetts oyster brands around the world.

The last unique appellation within Massachusetts is the
island of Martha's Vineyard, which is a region unlike any other

---

* Augustus Gould, *Invertebrata of Massachusetts* (Cambridge, MA: Folsom,
Wells, and Thurston, 1841), 136.

in Massachusetts or the world. Known by the Native American Wampanoag Tribe as Noepe (land amid the streams), the 124 miles of shoreline mark the furthest southern extent of the last ice age and boast iconic clay cliff formations over one hundred feet above sea level, a northern rocky shoreline, sandy east and south shores, and multiple brackish salt ponds fed by freshwater streams meandering across the lush forests of oak, maple, beech, and pine trees. Native American oyster middens are found along the island coast, and the highest elevation on the island is over three hundred feet and consists of pure glacial deposits. Ancient megalodon teeth, whale bones, and prehistoric fossils are found regularly within the coarse-grained stratified glacial deposits; however, it is the waters that make the island truly unique. Surrounded by the swift, cold-flowing waters of the Vineyard Sound, lobster-, eel-, and striped-bass-flavored saltwater tides flush into the ponds twice a day to provide a Wellfleet-like brininess to a clean, minerally, sweet, land-influenced merroir.

Unlike other regions in New England, Martha's Vineyard oysters are grown on entirely subtidal farms due to the extreme elevation. Farms are underwater twenty-four hours a day and incorporate tidal upwellers and bottom cages for growth. Utilizing small boats with cranes, the highest-quality farmers on the island periodically air-dry their gear with oysters inside to mimic the naturally occurring low tides, keep the oysters clean, and build up the adductor muscle inside each oyster for a sweeter finish. Farmers take great care to power-wash the product clean after harvest, producing very high-quality oysters served raw on the half shell. A Martha's Vineyard oyster generally fetches two to three times as much in the market as other New England oysters, making them the most expensive on the East Coast. Katama Bay in Edgartown, the third-largest producing region in the state, contains more than thirteen farms alone. In all, more than twenty farms inhabit the Vineyard, with different brands and merroir ranging from the open-ocean Cottage City Oysters farm to the salt pond farms like Signature Oysters in Katama Bay and Spearpoint Oysters in Menemsha Pond. Oyster farm tours are a popular tourist attraction, as are the numerous raw bars found around the island. Martha's Vineyard oysters are now shipped around the world, and honestly, they are all amazing.

Rhode Island

SHARING MUCH OF the same geological history as Massachusetts, Rhode Island differs in that the state produces the least salty of New England oysters. The Rhode Island oyster is a bay oyster, coming from the brackish waters of Narragansett Bay or the farther-inland brackish ponds around the coast. In a maze of marshland and eel grass, Rhode Island oysters are grown in bottom-culture bags or cages within the estuaries and typically carry a sweet, mild, and delicate merroir sensation.

One of the most productive salt ponds in Rhode Island is Point Judith Pond, a region home to a handful of the most famous shellfish farms in the state. Matunuck Oysters is the leader of the pack, having created an oyster experience dubbed "pond to plate," where guests can savor the fresh oyster merroir while gazing over the farm from the comfort of their restaurant seat. Considered one of the best oyster dining experiences in the world, Matunuck Oyster Bar not only sources the oysters from a nearby seven-acre farm in Potter Pond, but also sources the majority of plants and vegetables from their private farm just a mile down the road. Farm tours and educational sessions are held at the restaurant, and the company has their own hatchery as well, completing the vertical integration business model.

Owner Perry Raso has become a recent pioneer and spokesman for the industry, giving TED Talks about global food demands and the ability of shellfish to create a more sustainable future for the Earth.

Maine

WHILE THE GEOLOGICAL history of Maine is much the same as the rest of New England, the shoreline is entirely different. More akin to the West Coast of America, Maine coastlines harbor large bedrock outcroppings covered in feet of muddy clay silt. Pine trees and evergreen conifers fill every inch of land available, while enormous rivers drain the snowmelt from mountainous hills into the deep, frigid Canadian-like ocean.

The most famous region of Maine is "Down East" Acadia National Park with its landscape of waterfalls, bald eagles, moose, granite outcroppings, and untouched wilderness. The famous Bar Harbor is visited by cruise ships and tourists year-round.

Just south of Down East Maine is the Damariscotta River, or the "river of many fishes," as the Abenaki Native Americans called it, a nineteen-mile-long tidal river in Lincoln, Maine, that empties into the Atlantic Ocean. Shell middens dating back as far as 2,500 years ago sprinkle the riverbanks as testament to the productivity of the river waters.

Today, the river not only hosts a handful of oyster farms, like Glidden Point, who grow their oysters in floating cages and finish them on the river bottom, but it is also home to one of the largest shellfish hatcheries in New England: Mook Sea Farm, who maintains an advanced research and experimentation operation in addition to hatchery and farm operations. Mook's hatchery is one of the most state-of-the-art facilities in the world, utilizing off-grid energy systems and climate change mitigation designs.

Next door to the Damariscotta River is Muscongus Bay, where Muscongus Bay Aquaculture resides, another one of the largest shellfish hatcheries in New England. Together, more than half a billion oysters are created annually between these two hatcheries, which has become a lifeline to the entire Maine and Massachusetts aquaculture industries. Without these two hatcheries, much of the oyster seed grown in Maine and Massachusetts would not exist.

Maine produces some of the finest northern latitude oysters to be found. Mostly utilizing float gear, Maine oysters take upwards of four years to grow and typically contain a round, smooth, extremely hard, thick shell with plump, salty meats.

Connecticut

Over two hundred thousand bushels of oysters are harvested annually from the seventy thousand acres of shellfish farms in Connecticut. The majority of this production is done with the old-fashioned bottom-planting and dredge harvest method. Per linear mile, Connecticut's shoreline has more tributary rivers than any other region in the United States, making the environment one of the richest and most productive in the world for oysters.

Considered since the 1800s to be the oyster capital of the world, Connecticut has a long-standing history of oyster farming that predates the formation of the United States. Prior to the existence of independent states, fishermen were able to acquire "franchises" directly from the king of England. Purchased in the 1700s, these franchises are still valid today and provide the ownership rights to the underwater acreage. Owners pay property tax on the acreage, but also outright own their beds, making them the exception to New England "public" leases.

One of these early adapters was the Tallmadge Brothers Oyster Company, based in Norwalk, Connecticut, which was one of the first to begin converting sail-powered dredging boats to steam-powered dredgers. By 1880, Norwalk had the largest fleet of steam-powered oyster boats in the world. In the early 1900s, as Long Island oyster businesses started their long period of production decline, the owner of Tallmadge Brothers invited two fraternal twins named Hillard and Norman Bloom to come work for him. By the 1960s, the brothers were buying up most of the failed oyster companies of Long Island and their leases, and in 1967 they bought Tallmadge Brothers as well. By consolidating the shell companies under the Tallmadge Brothers Oyster Company, the Blooms were able to own the lion's share of the Long Island market and became synonymous with the "Blue Point" oyster brand.

"Blue Points" were a trade name first used by Joseph Avery, a veteran of the War of 1812, who returned to his home of Blue Point, Long Island, with a shipload of Chesapeake Bay oysters and planted them on a lease he had acquired in the Great South Bay. The brand became so popular that from 1817 on, any oyster that came from Great South Bay was labeled as a "Blue Point."

With the market cornered, the Bloom brothers bought the Bivalve Packing Company in 1972, which included the entire town of Bivalve, New Jersey. After the death of Hillard Bloom in 2001, the Tallmadge Brothers company and some twenty-two thousand acres of oyster beds were split up into two companies: Hillard Bloom Shellfish, Inc., and Norm Bloom & Son.

Run by the Bloom heirs, these two companies were responsible for the majority of oysters produced in Long Island Sound until 2019 and Covid, which brought the Hillard Bloom Shellfish company to the breaking point. Faced with diminished markets and an aging workforce, the family made the decision to close shop and sell off all assets.

The other half of Tallmadge Brothers, Norm Bloom & Son, got its start in 1994 and continues operations to this day. They are the largest producer of the famous Copps Island oyster brand.

With over twenty-two thousand acres in Long Island Sound, from Greenwich to New Haven, Copps Island Oysters utilizes the bottom-dredging method of farming, in which seed oysters are cast onto the bottom sediment of the lease site and harvested two to three years later with a large net dragged by a fishing vessel. The oysters are the closest thing to a wild-caught product possible and are utilized in every way, from being used in cooked dishes to eaten raw on the half shell. The sandy shallow seas and nutrient-rich brackish waters make them the go-to merroir for Connecticut.

GreenWave, perhaps the most famous New England aquaculture organization today, started as an oyster farm (Thimble Island Oysters) in Connecticut and quickly became the U.S.'s nonprofit leader in seaweed farming and regenerative farm systems after owner Bren Smith experimented with growing sugar kelp on the farm sometime around 2012. Cofounded with Emily Stengel, GreenWave has catapulted regenerative farm systems (which grow crops such as seaweed and oysters together) into the mainstream spotlight, appearing in such programs and publications as BBC Earth, TED Talks, *TIME* magazine, *60 Minutes*, *Rolling Stone*, the *New Yorker*, and more. Like most of us in the industry, Smith believes in a future in which ocean farms scatter the coastline of every major city around the world, providing food, jobs, cleaner waters, ecosystem restoration, and climate change solutions. Green-Wave's mission is to help train and support ten thousand farmers, with the larger goal of growing and scaling regenerative ocean farming around the world.

The Chesapeake Bay

THE CHESAPEAKE IS about the farthest location south along the East Coast of the U.S. where oysters are predominantly eaten raw on the half shell for their flavor and merroir. Obviously, there are exceptions. Some North and South Carolina regions are known for producing extremely sought-after blue-green gill oysters. However, the farther south you travel, the more bacteria-laden the warmer waters become, and the lack of free amino acids in the low-salinity waters create a blandness in the oyster flavor.

While not held in as high esteem as the salty oysters of New England, the Chesapeake Bay oyster is historically one of the most harvested and adored oysters from America, as we learned in the history of the oyster wars. Historically, the mild, soft texture of the Chesapeake Bay oyster was canned and shipped around the country, providing West Coast settlers with a small taste of their East Coast memories.

The geology and formation of the Chesapeake varies greatly from the New England coastline, giving Chesapeake oysters their own unique blend of ancient minerals and influence.

During the Eocene Epoch (fifty-six to thirty-four million years ago), ocean levels were about 150 meters higher than present day, which put the coastline at the foot of the Appalachian Mountains, rendering the modern areas of Richmond, Washington, DC, and Baltimore completely underwater. Extending from this ancient coastline to the continental shelf was most likely a very large oyster reef, as noted by the buried calcium carbonate found today. Trillions of oysters living along the shoreline were suddenly wiped off the planet as the Eocene Epoch came to a screeching halt about thirty-five million years ago when a three-to-five-kilometer bolide (comet or asteroid) plunged into the Earth's atmosphere and pulverized a giant crater into the present-day Chesapeake Bay region.

Located about two hundred kilometers south of Washington, DC, and buried some three hundred to five hundred meters beneath the Chesapeake Bay, the bolide crater is twice the size of Rhode Island (eighty-five kilometers in diameter), deeper than the

Grand Canyon (1.3 kilometers deep), and the sixth-largest crater known on the planet.

In the thirty-five million years that followed, sediment filled in and around the crater, and by eighteen thousand years ago the oceans had fallen two hundred meters lower than present day, leaving the Chesapeake Bay region a grassland of rivers flowing with nutrients provided by the Appalachian Mountains. The crater depression became a natural destination for the James, York, Rappahannock, and Susquehanna rivers, and around ten thousand years ago, as the enormous glacial ice sheets began to melt, ocean levels rose and the river valley formed the Chesapeake Bay.

The Chesapeake Bay is now the end point for over 150 rivers and streams and forms the largest estuary in the United States. Once considered the most productive oyster grounds in the world, the Chesapeake Bay's oyster harvest has now been reduced to less than 1 percent of the traditional harvest average.

The most important organization in the region is the Chesapeake Bay Foundation (CBF). Founded in 1966, and now with offices in Maryland, Virginia, Pennsylvania, and DC, plus fifteen field centers, the organization leads the way in restoring the bay and nearby rivers and streams. With a motto of "Save the Bay," it's no surprise that oyster reef restoration is one of their flagship programs.

In 2010, CBF settled a lawsuit with the EPA that provided science-based, enforceable limits on the amount of pollution entering the Chesapeake watershed in order to remove the bay from the "dirty waters" federal list. To help meet the environmental goals, CBF launched the Chesapeake Oyster Alliance, an organization to help carry forward their oyster restoration efforts, in 2018.

Now a coalition of more than ninety nonprofit, private, state, and federal agencies, the Alliance recorded its 4.47 billionth oyster planted in the fall of 2022, making it nearly halfway toward completing the ambitious goal of planting ten billion new oysters in the bay by 2025.

One of the major efforts of the Alliance is to increase aquaculture farms within the bay. For decades, a wild oyster fishery culture has consumed the region, frowning upon individuals who try to "own" a part of the bay as a private aquaculture farmer. This mentality kept most aquaculture hopefuls from being able to obtain local permits, and for those who did start farms, theft and vandalism took a large toll. However, within the last decade the tide has turned. Education, science, and market demand have changed the perceptions of the region, and gradually aquaculture farms have been on the rise, creating new products and brands and helping to put the Chesapeake oyster back on the map.

The Chesapeake hosts a wide range of microclimate merroir regions within the bay, and many farms utilize a bottom-growing system and trawls, mimicking a wild oyster fishery. The most recognizable outfit within the Chesapeake is the Rappahannock Oyster Company, founded by the Croxton cousins in 2002. Utilizing lease sites once owned by their grandfather (dating back to 1899), the cousins decided to experiment with a bottom cage system, placing the oysters in bags that sit in a cage on the bottom sediment. As product began to come to market, the Croxtons set up tastings with high-end chefs around the country and were met with fanfare—a Chesapeake oyster to celebrate! Their success has led the Rappahannock oyster to be one of the most recognizable brands out of the Chesapeake, now with five oyster bars ranked as some of the best around the country and a distribution system to back it.

The American West Coast

THE MAJORITY OF the Pacific Ocean is surrounded by a rim of volcanoes called the Ring of Fire, which extends some twenty-five thousand miles along the coasts of South America, North America, Russia, and some of the islands in the western Pacific Ocean.

About thirty-six million years old, the portion of the Ring of Fire that inhabits the West Coast of America is called the Cascade Volcanic Arc, which spans from Northern California to southwestern British Columbia and includes a band of thousands of very small, short-lived volcanoes that have built a platform of lava and volcanic debris. Rising above this platform are the larger volcanoes, such as the famous Mount Rainier and Mount St. Helens.

This volcanic soil mixed with glacial activity has shaped the Northwest landscape. Looking very much like Maine, evergreen conifer trees blanket the land as rivers carrying volcanic mountain minerals deposit into shallow tides and muddy bottom sediment fed by the upwelling nutrients of the cold Pacific Ocean currents.

It's no secret that clams grow best in the Northwest muddy tidal flats, with the tiny half-dollar-sized Olympia oyster being able to take hold in only the most sheltered bays along the coast. Unfortunately for the oysters, the West Coast of America has very few inlets available for growth. San Francisco is the only real bay within California suitable to the oyster, which was quickly discovered and harvested to near extinction by the 1860s.

Oyster exploitation moved north into the Puget Sound–Seattle region, and startup operations quickly began harvest of the native Olympia oysters to meet the increasing demand. Olympia seed was shipped from these regions back to San Francisco Bay; however, the heavy lifting of the industry remained in Puget Sound.

In 1875, transcontinental trade for Eastern oyster seed was established, and shipments of market-size *Crassostrea virginica* oysters were transported by train in barrels of sawdust and ice and planted into the San Francisco Bay in attempt to propagate them on the West Coast. The East Coast oysters grew, but the waters never reached warm enough temperatures to allow for spawns. Another source of oysters was desperately desired.

No other company encapsulates the story of the American West Coast oyster better than Taylor Shellfish, a fifth-generation company and the largest shellfish business in the United States, with six hundred employees, nine thousand acres of shellfish beds in production, and operations in the U.S., Canada, Hong Kong, and Fiji.

Started in Puget Sound as the Olympia Oyster Company by James Waldrip in 1890, the business helped fulfill the West Coast oyster demand. In the 1920s, Waldrip's operation was one of the first to import the Japanese Pacific oyster to the West Coast. Seed shipments from Japan kept the industry alive, and as the Pacific oyster began to reproduce on its own in the new American region, it quickly dominated the landscape, outcompeting the native Olympia oyster in growth rates, size, and durability.

Waldrip's son (James Waldrip Jr.) kept the business afloat during the Great Depression; however, by the 1940s, overharvest and sawmill pollution caused the native Olympia oyster population to crash. Interestingly, the Pacific oysters were able to survive and became the de facto oyster on the West Coast moving forward.

During World War II, seed imports from Japan were halted, and the crumbling industry began to confront the inevitable need for a sustainable seed option. Wild Pacific oyster seed was collected, and the struggling industry was kept alive. Waldrip's tideland deeds were kept active, but it wasn't until the 1950s that grandson Justin Taylor decided to begin farming shellfish full time.

Oyster seed was again imported from Japan, with new varieties like the Kumamoto being introduced. However, seed prices had become very expensive by the 1970s, and in response, an Oregon shellfish hatchery began offering clam and oyster seed to growers. By the 1980s, Taylor Shellfish decided to build their own shellfish hatchery and began extensive breeding programs. Their products today include Manila clams, geoducks, Mediterranean blue mussels, Pacific oysters, Kumamoto oysters, *Crassostrea virginica* oysters, and Olympia oysters. This is the only location on Earth where four distinct varieties of oysters can be consumed that are grown in the same region. With restaurants up and down the coast, Taylor Shellfish *is* the West Coast oyster merroir factory. The company is now exploring different growing methods to shape new products and to learn the nuances that create unique merroir flavors.

While not as old as Taylor Shellfish, the Hog Island Oyster Company, founded in 1982 in Tomales Bay (a small inlet just north of San Francisco), is well deserving of mention. Founded by two marine biologists and utilizing high-tech farm methods from around the world, the company has been able to not only capture the merroir of Tomales Bay into beautifully crafted products, but they have also led the movement in farm-to-table offerings.

Hog Island Oyster Company has without a doubt the best oyster bars on the West Coast, with restaurants in Marshall, San Francisco, Napa, and Larkspur. They are pushing the industry into new products like red abalone, seaweed, locally sourced fish, and a variety of tinned fish.

WHERE DO WE GO
FROM HERE?

IT'S IMPORTANT TO POINT OUT THAT THESE ARE
not the only oyster appellations around the world. In fact,
this is just the tip of the oyster production iceberg. Unlike
wine grapes, which require specific growing zones and climates,
oysters have the flexibility to grow in almost any location
around the world, enabling an amazing array of flavors and farm
differences.

As oyster culture continues to expand, additional sought-
after regions will be recognized, and eventually the nuances in these
regions will become apparent. Locations such as Ireland, the Netherlands, Portugal, Britain, Italy, Croatia, Canada, Mexico, Alaska,
Peru, and Africa are developing what will be highly desirable oyster
appellations of the future.

We are at the dawn of a new era for culinary enthusiasts. Like
being witness to the founding of the wine industry, consumers

today are at the forefront of trying to describe these new flavor profiles and nuances found in oysters around the world.

Will it take centuries to build out these ideas as it did with wine and the Benedictine and Cistercian monks, or will we be able to use modern-day techniques and terroir as a model to speed up the task?

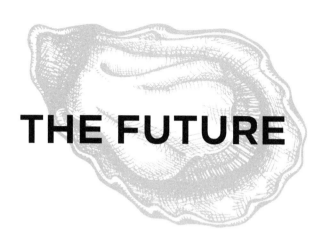

THE FUTURE

THE FUTURE
OF FOOD

HOPEFULLY BY NOW YOU HAVE AN UNDER-
standing of the rich history of oysters and how the current industry operates. Before we map a trajectory of
where the oyster farming industry is going, we need to examine
the problems currently facing the Earth and our food production
system. By understanding these problems, we begin to have a better
appreciation for oysters and other shellfish, and it becomes apparent very quickly that they offer solutions to many of our future
food challenges.

To look to the future of the oyster industry, we must first
address the elephant in the room—climate change and our failing industrial agriculture complex. The failures of our agriculture
system are well documented, and the pressure that climate change
will place upon this failing system is undoubtedly the straw that
will break the camel's back. Regardless of how or why the climate is

changing, the magnitude of the climate shift will change the food industry on Earth as we know it today. Why these facts are not being discussed publicly is disturbing. Perhaps the issues seem too large or complex to solve. Perhaps it's easier to put our heads in the sand and hope others solve the problems. Nevertheless, it's time for some real talk. Yes, the facts are depressing and an unfortunate reality, but at the end of this book we will highlight victories, successful works in progress, and ways you can pitch in. We will get through it, like humanity always does.

The contribution of fossil fuels into the atmosphere by human activity is helping to accelerate climate change; however, the science tells us that even if we stop all fossil fuel usage today, the Earth has already crossed the threshold of returning to any kind of climate we've known in the last one hundred thousand years.

Change is nothing new. The Earth has been changing since the day it was created. Instead of measuring how human activity is changing the climate, we should change the narrative to measure how much human activity is helping to stabilize the environment.

Utilizing carbon sequestration technologies and reducing our fossil fuel usage will help stabilize the climate and ultimately put the human race on the path of global climate control or climate stabilization (an agreed-upon global climate). However, before we get anywhere close to climate control, the polar regions will continue to melt, the oceans will continue to rise, the ocean currents will continue to change, and in response, weather patterns around the world will change with them. Climate change is inevitable, and luckily, humans are great at adapting.

When compared to historical practices, it is apparent that there is a great mindset shift taking place within humanity over the last few decades. Finally, the human species has realized that Earth's resources are finite. The undeniability of this realization has only occurred in recent decades with the advent of the internet, which has opened communication on a global scale. For the first time in human history, we have real-time communication and data sharing around the world—and not just among the super wealthy and powerful. Common citizens with access to the internet can now globally communicate, share data, and educate others on whatever topic they desire. Food commodity prices and production numbers are shared. Freshwater aquifer health and usage is monitored and shared. The age of global data collection and sharing is here, and with it comes the reality of our resource usage footprint.

The Covid-19 pandemic was the first worldwide phenomenon of the "global connected age" to expose civilization to what a response to a global problem could look like—complete with all the successes and failures. In response to Covid, for the first time in history, global resources were shared in real time. Information, both true and false, spewed rampantly around the world within minutes of each new revelation. Science took the reins as knowledge about the virus was shared between countries, vaccines were developed, and supplies were shifted around the globe to where they were needed most, all while fear and cynicism reached all-time highs. For the human species, all the flaws and triumphs were on full display, and yet here we are, together, as a unified species; we survived and adapted.

Climate change will inevitably make Covid look like child's play. In a future with unstable weather patterns and climate shifts, food production will become unreliable—a problem that humanity has yet to face. Areas today that are highly productive will become void of necessary resources. Think of the Dust Bowl of the 1930s, or the more recent breakdown in global shipping during Covid. Climate change will inevitably break down global commodity shipping and rearrange food productivity regions, forcing grocery shelves to switch from scarce global supplies to more locally sourced goods with higher prices. Fear and cynicism will again reach all-time highs as superstorms demolish neighborhoods, countries go to war over food and water supplies, and the power grid struggles to maintain stability. Yet we will continue to prosper.

Until humans learn to control the climate, food production will remain the number one focus of civilization moving forward. My perspective as a farmer, the Farm Bureau president of my local chapter, and a board member for multiple agriculture organizations has opened my eyes to the crisis we are facing. As someone who works with the environment and talks about these issues with other farmers, who is witnessing animal population shifts firsthand, who is highly invested in understanding what the future will be so that I and others can adapt our farms accordingly to ensure they remain productive, I am finding it increasingly obvious that the industrial agriculture "experiment" of the last sixty years is on the verge of total collapse, and I'm not the only farmer who thinks this way. The vast majority of farmers you talk to today are beginning to grasp what the future of agriculture will look like, and it's not pretty. One thing is for sure, without ocean farming, our species will not continue to thrive.

While it's anyone's guess what the future may be, there is one priority going forward: we must focus our attention on the best possible use of our existing resources.

Consider the Earth as a spaceship locked in orbit around the Sun. When launched, the Earth was packed full of a finite number of resources to sustain life, such as fresh water, phosphorus, lithium, and other trace minerals. The human race has consumed these resources at a speed much faster than they can replenish, and the Earth is now reaching the point where demand for these scarce resources is overwhelming the supply.

Rare earth minerals, used in modern medical devices, electric cars, and renewable technology, are naturally occurring resources present in only very small quantities in the Earth's crust. There are five that modern medicine and renewable technology depend on today—tantalum, silver, lithium, gallium, and indium—and all five are expected to run out in fewer than one hundred years if the current consumption rate continues.

To exacerbate the problem, demand for these rare minerals is rising annually. Perhaps our need for these materials will lead us to mine the deep oceans, the Moon, asteroids, and planets beyond Earth. Fortunately, humanity will not die if these resources run out. As important as they are, we could argue that they are nonessential to our survival.

Our food production industries, on the other hand, also face diminishing raw materials, and if these resources are not renewed, humanity *will* die out.

For humans, the two most precious resources on Earth are land and fresh water, as both are required for survival. Land for houses and food, and water for our nourishment. Our food production systems also require land—specifically fertile soil—and fresh water, both of which are finite resources. As our population continues to grow, additional land and fresh water will be needed for human consumption, which will put additional pressure on our food production systems to expand, which will require even more land and fresh water for food production! In essence, we are competing with our food system for the same resources we require for our own species' growth and survival.

This trend cannot continue. Our food production systems must get smarter. Over nine billion humans are expected on the planet by 2050. By 2100, the estimated human population will be around eleven billion people. The global population has more than quadrupled in the last one hundred years, with an estimated eighty-two million additional people now added to the planet every year. The Food and Agriculture Organization (FAO) of the United Nations estimates that

> *Nearly all of this population increase will occur in developing countries. Urbanization will continue at an accelerated pace, and about 70 percent of the world's population will be urban (compared to 49 percent today). Income levels will be many multiples of what they are now. In order to feed this larger, more urban and richer population, food production (net of food used for biofuels) must increase by 70 percent. Annual cereal production will need to rise to about 3 billion tonnes from 2.1 billion today and annual meat production will need to rise by over 200 million tonnes to reach 470 million tonnes.*

In looking at human history, it is easy to forget where we came from and where we might be going. Humans were hunter-gatherers up until about twelve thousand years ago, when we began to farm the Levant. The advent of agriculture created profound changes to civilization and is responsible for the modern society we have today.

I think of fishermen as the hunter-gatherers of the oceans, tracking schools of prey and harvesting what the wild can provide. With aquaculture, we are at the beginning of our ocean farming history. Like the invention of agriculture, aquaculture will have profound implications for our food systems and our greater society

---

* FAO, *How to Feed the World in 2050* (FAO, 2009), fao.org/fileadmin/templates/wsfs/docs/expert_paper/How_to_Feed_the_World_in_2050.pdf.

moving forward. Aquaculture will likely be the number one influence on our food systems for centuries to come.

What fascinates me most about the development of ocean farming is that we are building this pillar of society now, in a time of great awakening for our species, in which we have begun to think about the planet and the environment, not just ourselves, as we have done in the past.

Seafood inherently uses the fewest finite resources for production, requiring minimal land or fresh water. With aquaculture, some fish farming is land-based and can utilize freshwater inputs in the form of fish feed; however, marine aquaculture systems that farm in the ocean waters, like shellfish and seaweed farming, are the most sustainable protein farming systems possible on Earth, requiring zero land, zero fresh water, and zero feed inputs.

As cities expand exponentially around the world, local food production is becoming more important. The fewer resources our foods require, the more sustainable they become, yet today the average distance processed foods travel in the United States is roughly 1,300 miles, while fresh produce travels about 1,500 miles before it reaches the consumer.[*]

Covid showed the fragility of the transportation sector of our global food system and ultimately broke down some of this reliance. Products that were transported around the world suddenly disappeared from store shelves, which in turn provided opportunistic domestic producers to fill the gaps. Consumers and distributors were forced to either seek out a local producer or go without some products. To aid the distributors, some government

---

[*] Holly Hill, "Food Miles: Background and Marketing," National Center for Appropriate Technology, 2008, attra.ncat.org/publication/food-miles-background-and-marketing.

regulations that hindered the sale of locally produced foods were rescinded, returning markets back to the age of pre-global agriculture. Seafood was one of the largest benefactors of these regulation rollbacks, allowing fishermen, for the first time in decades, to sell their catches directly off the boat to the public.

In the aftermath of Covid we have seen a returned focus to local food systems and the necessity of domestic production. Food security is a rising concern for all communities and will continue to be tested as climate change accelerates. Freshwater resources, transportation infrastructure, land access, energy production, and storage systems will ultimately break down in the wake of an unstable climate. Covid provided a glimpse into our societal weaknesses, and many organizations are now focused on strengthening these chinks in our armor.

As our population continues to grow, experts agree that food production will need to increase to provide for these new mouths, but our input resources and farmable land continue to decrease annually. Some studies even go so far as to argue that we only have sixty years of nutrients left in our soils if we continue down the chosen agricultural path.* We will need to produce more food in the next fifty years than has been produced in the last twelve thousand years. According to the UN,

> *Water, food and energy form a nexus at the heart of sustainable development. Agriculture is the largest consumer of the world's freshwater resources, and water is used to produce most forms of energy. Demand for all three is increasing rap-*

---

* Chris Arsenault, "Only 60 Years of Farming Left If Soil Degradation Continues," *Scientific American*, December 5, 2014, scientificamerican.com/article/only-60-years-of-farming-left-if-soil-degradation-continues.

idly. To withstand current and future pressures, governments must ensure integrated and sustainable management of water, food and energy to balance the needs of people, nature and the economy.[*]

We need to manage our finite resources more wisely, limit our environmental impacts, and repair the ecosystems we have destroyed to date. Aquaculture, specifically shellfish and seaweed farming, is our best bet at meeting these demands. Aquaculture can help reduce seafood's carbon footprint and provide protein- and nutrient-rich foods without the use of additional land or fresh water, while providing local jobs and strengthening local economies.

Nearly every coastal country on the planet has significant potential to farm the oceans, yet this opportunity has remained out of reach due to pollution and lack of technological breakthroughs in species seed sources, infrastructure materials, and access—until now. The recent breakthroughs in aquaculture over the last decade have created an opportunity for the industry to expand into every major coastline around the world, without a moment to spare.

In January of 2023, a scientific paper entitled "Carbon Sequestration via Shellfish Farming: A Potential Negative Emissions Technology," by Jing-Chun Feng, Liwei Sun, and Jinyue Yan, presented unequivocal data to demonstrate that if shellfish farming were expanded into all possible locations around the globe, a total of 17.63 percent (5.64 Gt CO_2-eq) of all 2020 emissions could be sequestered annually. This is equivalent to removing all carbon emissions from India. The report takes into account the sequestration of carbon through oyster meats, feces, and shells, and

---

[*] United Nations, "Water, Food and Energy," unwater.org/water-facts/water-food-and-energy.

concludes that shellfish farming creates a unique opportunity to reduce our carbon footprint while also increasing our food supply. No other food or technology can come close to the same results. The paper ends with a call to begin implementation of shellfish farms around the world in all possible locations immediately.

Amazingly, the report does not consider the prospect of utilizing shellfish for animal feeds, which can further reduce resource inputs by replacing the resource-intense ingredients already found in these feeds. By using shellfish to form the "feed" base of the industrial food chain, we can lower the resource use of traditional farm animals while restoring naturally produced nutrients back into the arable soils.

The opportunity today, to design a future food pillar of our society, presents the greatest optimism for the environment and future of the human race, but only if we choose to implement these ocean farms wisely.

To meet the growing demand for food we need to redesign the system, as the current status quo will not work. One of the most serious challenges for the current generation is to provide land, water, food, and energy in a sustainable way for a large world population. It will not be an easy task.

LACK

OF SPACE

LAND IS LIMITED. ARABLE LAND IS EXCEPTION-
ally limited. Twenty-nine percent of the Earth's surface is
land and 71 percent of that land is habitable. Half of the
world's habitable land is already used for agriculture! If we com-
bine pastures used for grazing with land used to grow crops for animal feed, livestock accounts for 77 percent of all global farmland.
While livestock takes up most of the world's agricultural land, it
only produces 18 percent of the world's calories and 37 percent
of total protein. In essence, we have dedicated half of the land on
Earth to livestock production and the support thereof, for it to
only produce a little more than one-third of our total protein. This
is a losing proposition. On the other hand, farm crops use about 23
percent of the available agricultural land and produce 83 percent

of the global calorie supply and 63 percent of the protein.* This
represents an amazing use of space!

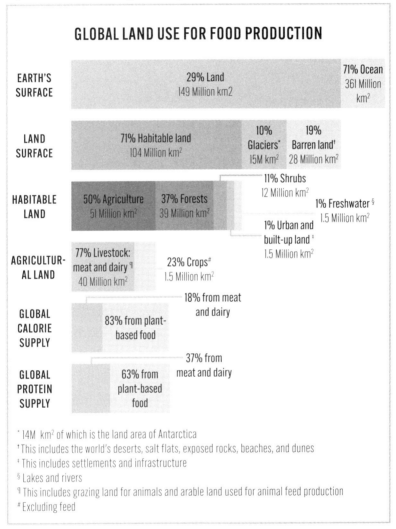

Data source: UN Food and Agriculture Organization

---

* Hannah Ritchie, "50% of All Land in the World Is Used to Produce Food,"
World Economic Forum, December 11, 2019, weforum.org/agenda/2019/12/
agriculture-habitable-land.

In total, more than 80 percent of human protein consumption is created on land. While this might sound impressive, it doesn't come close to meeting the needs of today's population. Nine percent of the world population (roughly 697 million people) globally are severely food insecure and undernourished, meaning they have a caloric intake below minimum energy requirement. One in four people globally (1.9 billion) are moderately or severely food insecure.[*] The food production systems of today can't even meet the present global food demand.

To make matters worse, one-third of the planet's land is already severely degraded, and fertile soil is being lost at the rate of twenty-four billion tonnes a year, according to the Global Land Outlook.[†]

A 1993 statement issued jointly by academies of science from all over the world provided an early warning:

The growth of population over the last half century was for a time matched by similar world-wide increases in utilizable resources. However, in the last decade food production from both land and sea has declined relative to population growth. The area of agricultural land has shrunk, both through soil erosion and reduced possibilities of irrigation. The availability of fresh water is already a constraint in some countries. These are warnings that the earth is finite, and that natural systems are being pushed ever closer to their limits.[‡]

---

* Hannah Ritchie, "How Is Food Insecurity Measured?" Our World in Data, April 27, 2023, ourworldindata.org/food-insecurity.

† United Nations Convention to Combat Desertification, *The Global Land Outlook*, Bonn, Germany, 2017, unccd.int/sites/default/files /documents/2017-09/GLO_Full_Report_low_res.pdf.

‡ National Academy of Sciences, National Academy of Engineering, and Institute of Medicine, *Population Summit of the World's Scientific Academies* (Washington, DC: The National Academies Press, 1993).

With 29 percent of the Earth being land, the remaining 71 percent is covered by ocean, yet the ocean produces just 2 percent of our food. It hasn't always been this way. Traditionally, our species relied heavily on ocean resources for sustenance, but learning the hard way, we have recently discovered that our ocean resources are not infinite. Eighty percent of wild fish stocks are currently overfished or depleted. Globally, fish and other types of seafood now comprise just 6 percent of the protein consumed by humans. Learning to properly manage and rebuild the stocks will take decades, if it's even possible at all. Unfortunately, wild catch seafood is unlikely to contribute to our growing food production demands.

Our farmlands don't seem to offer any optimism either. In 2022, the nonprofit American Farmland Trust compiled a report entitled *Farms Under Threat*, which identified that more than 1.1 million farmland acres were lost to sprawling development from 2001 to 2016. Utilizing forward-looking trends and data, the report suggests that by 2040 in a "business as usual" future, more than 18.4 million acres (the size of South Carolina) will be converted from farmland to developments, while in a "runaway" development scenario, the losses could climb to as many as 24.4 million acres lost.

With so much land already dedicated to food production, it's easy to see how expanding development directly competes with our food production systems. We are quickly running out of land to expand! And while we may be able to expand our production efficiencies and squeeze a little more juice out of the fruit, these advances are unlikely to meet the upcoming global food needs.

We may be near the end of our land-based food production capabilities; however, there is clearly a golden opportunity within our oceans that we have yet to seize. If the human species wishes to continue to grow, we must first learn to farm the oceans.

Taking this idea literally, the Italians have found an interesting solution to make more arable land available: they have begun to grow the plants underwater.

Off the coast of Noli, Italy, just southwest of Genoa, cofounder Luca Gamberini created an underwater garden dubbed "Nemo's Garden," the world's first underwater cultivation system of terrestrial plants.

The garden has been growing for ten years strong and consists of multiple biospheres, two meters in size, floating six to ten meters below the ocean surface. Chained to the ocean bottom, the biodome pockets of air act like perfect isolated greenhouses, separated from pathogens and pests and with temperatures regulated by the relatively stable ocean waters.

Sunlight penetrates down through the ocean waters and into the plants growing inside the biodomes, though supplemental light is used when needed. The plants grow in a hydroponic solution of nutrients, while naturally occurring condensation within the bubble also helps keep them growing. Harvesting occurs with the aid of a diver, who enters the biodome from below and "surfaces" inside the container. The diver then harvests the plants, places them inside an airtight container, and releases them to the surface where a boat awaits their arrival.

Anywhere from seventy to one hundred tiny plants can fit within each biodome, and a 2020 study from Università di Pisa concluded that the basil from Nemo's Garden had more antioxidants and higher essential oil concentrations than standard terrestrial-grown basil.

Eventually the Nemo's Garden team plans to scale up to larger-sized biodomes and even has a pilot project small-scale version being installed in a cold-water quarry in Ohio.

As strange as growing plants in a bubble under the ocean seems, it may have to become reality as our status quo food production system attempts to keep up with a sprawling human population. As cities expand and fertile land becomes scarce, the development of the oceans as a food supplier begins to look much more practical.

Perhaps the future will contain enormous underwater green-houses full of orchards and fields of corn, where underwater farm cities harvest food for the land-dwelling humans. Undeniably, the future will consist of aquaculture farms scattering the coastline of every major coastal city on Earth, growing marine organisms such as shellfish, seaweeds, and finfish, in the most sustainable way possible.

FINITE
RESOURCES

THE ENVIRONMENTAL IMPACT OF FOOD AND agriculture on the planet cannot be underestimated. Twenty-six percent of global greenhouse gas emissions come from food production; 70 percent of global freshwater withdrawals are used for agriculture; 87 percent of global ocean and freshwater pollution is caused by agriculture; 94 percent of mammal biomass (excluding humans) on Earth is livestock. Of the twenty-eight thousand species evaluated to be threatened with extinction on the International Union for Conservation of Nature Red List, twenty-four thousand are threatened by agriculture and aquaculture. Our agricultural production systems not only fall short of demand, but they also destroy our planet in the process.

Almost 90 percent of global deforestation is caused by agricultural production. Worldwide, half of the deforestation is for cropland, while the other 40 percent is for livestock grazing. Between 2000 and 2018, the vast majority of deforestation took place in South American and Asian tropical rainforests, helping to exacerbate the effects of climate change.

As agriculture continues to transform the face of the planet, it also threatens to undermine our freshwater supply. Just 3 percent of the water on Earth is fresh water; however, two-thirds of that water is frozen in glaciers or otherwise unavailable for our use. This has resulted in 1.1 billion people worldwide lacking access to water, and 2.7 billion people finding water scarce for at least one month a year. At the current consumption rate, two-thirds of the world's population may face freshwater shortages by 2025.[*]

The Paris-based, international Organisation for Economic Co-operation and Development (OECD) lays it out fairly bluntly:

> *Irrigated agriculture remains the largest user of water globally, a trend encouraged by the fact that farmers in most countries do not pay for the full cost of the water they use. Agriculture irrigation accounts for 70% of water use worldwide and over 40% in many OECD countries. Intensive groundwater pumping for irrigation depletes aquifers and can lead to negative environmental externalities, causing significant economic impact on the sector and beyond. In addition, agriculture remains a major source of water pollution; agricultural fertilizer run-off, pesticide use and livestock effluents all contribute to the pollution of waterways and groundwater.[†]*

---

[*] "Water Scarcity," World Wildlife Fund, worldwildlife.org/threats/water-scarcity.

[†] "Water and Agriculture," OECD, oecd.org/agriculture/topics/water-and-agriculture.

In 2018, Cape Town faced the prospect of becoming the first major city in the world to run out of drinkable water. Record droughts between 2015 and 2018 saw the city on the brink of "Day Zero," the point when the municipal water supply would be shut off.

Thankfully, Day Zero never came, and since the 2018 emergency, the city's largest supplier of water, the Western Cape dams, had replenished to 64 percent by March 2024. Large rainfall amounts played a huge part in the recovery; however, the city's management strategies and the public water-saving efforts set the example for future crisis prevention.

The municipality's immediate response to the water crisis was to divert water from the agricultural sector to supply the city. Thirty thousand jobs in the agricultural sector were lost, and while the solution was not sustainable, it bought the city time to formulate a plan. It was the first time in recent history that a major city has had to choose between using water for growing food or for direct human needs.

Residents were limited to using just fifty liters (thirteen gallons) of water per household per day—for context, that is about eight toilet flushes. For comparison, the average American household uses about 138 gallons of water per day, or about sixty gallons per person.

To enforce the daily limits, Cape Town undertook intense water monitoring solutions and installed 250,000 smart meters at properties. These meters were equipped to turn the water off after the household daily usage limit had been reached. In addition, Cape Town increased the tariffs on water and enacted heavy fines for households that surpassed the daily limit.

Residents undertook massive gray water (wastewater from sinks and washing) recycling programs, reusing the water for toilet flushes and garden watering. Swimming pools were no longer allowed to be filled, car washes were banned, and garden watering became a nighttime activity to reduce evaporation.

Cape Town is not out of the woods. With relaxed rules, freshwater usage has begun to again creep higher. The Cape Town story may sound like a one-off problem, but it isn't hard to imagine a future where almost every major city in the world enacts some kind of freshwater limit on residents.

In a recent twist, Phoenix, Arizona, made headlines in June of 2023 for limiting all new construction in the city after it was proven that not enough fresh water existed for already approved construction projects. In March of 2024, Phoenix City Council approved a water conservation ordinance for "big water users" requiring water conservation plans for all new construction. Southwest states have ramped up water conservation initiatives in the last few years as the Colorado River has declined more than 20 percent in the last decade due to climate change and overuse, leaving the Lake Mead and Lake Powell dams on the brink of being unable to generate hydropower. The river provides drinking water to forty million people in seven states, thirty tribes, and Mexico.

Fresh water is a finite resource. How we use it will be the most important focus for the coming century, but it's not all doom and gloom. If we look at the problems as a glass half full, we can understand that it is our own doing that has painted us into the corner. If we made the mess, we can clean it.

A water tax on corporations and agriculture is not a new concept. Many in the food industry agree that in order to create a sustainable water supply program, you first must put a price on water—a concept that is just now being fleshed out. Leading the way in water pricing is California's Pajaro Valley, where more than thirty thousand acres of farmland and producers now pay up to $400 per acre-foot of water. The water tax is ultimately passed on to the consumer, raising prices of premium products like berries in the grocery store, while forcing producers of low-cost commodity crops like cotton, corn, and animal feeds to abandon the products all together.

A 2023 *New York Times* investigation determined that "many of the aquifers that supply 90 percent of the nation's water systems, and which have transformed vast stretches of America into some of the world's most bountiful farmland, are being severely depleted." The investigation revealed that "groundwater loss is hurting breadbasket states like Kansas, where the major aquifer beneath 2.6 million acres of land can no longer support industrial-scale agriculture. Corn yields have plummeted. If that decline were to spread, it could threaten America's status as a food superpower."[*]

The loss of groundwater is like a ticking time bomb waiting to explode, yet change doesn't come easy.

The Pajaro Valley Water Management Agency first installed meters to measure groundwater discharge in 1993 and began charging farmers thirty dollars per acre-foot to cover the costs. In response to severely overdrawn aquifers, the agency built a $6 million project to capture and divert rainwater into irrigation wells. A $20 million water recycling plant was built to clean five million gallons of sewage a day for irrigation. Now an $80 million system is being built to capture and store additional rainwater for irrigation. After more than two decades, the water pricing model is working, and word is getting out. Experts from China and Egypt now travel to the valley to study the system.

---

[*] Mira Rojanasakul, Christopher Flavelle, Blacki Migliozzi, and Eli Murray, "America Is Using Up Its Groundwater Like There's No Tomorrow," *New York Times*, August 28, 2023, nytimes.com/interactive/2023/08/28/climate/groundwater-drying-climate-change.html.

As our human population grows, we require more land and fresh water for consumption, and more food, which also demands more land and fresh water; however, there isn't any more to go around, and here lies the problem. How do we produce more food, while using fewer resources?

While this may sound impossible, it's actually the status quo throughout our history. Necessity breeds creativity. Our cavemen ancestors drove the wild resources around them to extinction through overhunting and overharvesting, and as a result, they were forced to turn to a new source of food production: land farming.

Following in these footsteps, modern fishermen have driven the wild stocks of the ocean to near extinction. We must bring agriculture to the oceans. Farming the ocean is the only plausible way to meet future food demands.

WORK IN PROGRESS

THE VERY RECENT SHIFT IN MINDSET ABOUT finite resources and climate change has prompted a transformation of the human race. Instead of consuming resources endlessly without thought of the effects of these actions (as we have done for the majority of history), we are now cognizant that Earth's resources have limitations and our actions have the ability to change the climate. The Earth spaceship is fragile, and we are now recognizing our impact. Suddenly, cries to obtain "net zero" (reaching a balance between the number of emissions introduced to the atmosphere and the number removed) are being touted by the likes of BP, Exxon, and most "progressive" companies on Earth.

Carbon capture, electric vehicles, and renewable power have begun to help stabilize our carbon footprint. Offshore mining technologies promise to increase access to newly found minerals, while recycling programs begin to return raw materials back into the production streams or mine old landfills for discarded treasures of the past.

Our food production industries seem to be the last to adapt. According to the FAO, livestock produce 7.1 gigatons of CO_2-equivalent per year, which is 14.5 percent of *all* global greenhouse gas emissions. The EPA says that agriculture, forestry, and other land use contribute about 24 percent of the total global greenhouse gas emissions.

Cattle (including beef and dairy cows, manure, and draft power) represents roughly 65 percent of those total emissions. About 44 percent of livestock emissions are in the form of methane (CH_4), a shorter-lived greenhouse gas when compared to carbon, but about twenty-five times more potent at trapping greenhouse gases in the atmosphere.

The FAO says that North America (including the U.S., Canada, Greenland, and Bermuda) accounts for just 1.2 percent of all global livestock emissions. This is mainly due to the success of their breeding programs. For example, in Mexico it takes up to five cows to produce the same amount of milk as one American cow, and in India, it takes up to twenty. The number of dairy cows in the U.S. has also dropped from twelve million in 1970 to just nine million, while producing double the milk! America has substantially reduced its cattle footprint due to increased efficiency and lower demand. Americans consumed about eighty pounds of beef per person in the 1970s, down to about fifty-seven pounds per person today. However, the consumption rate is only half the problem.

America exported less than 1 percent of its beef production in the 1970s, up to about 11 percent by 2018, and with increased beef demand and exports entering the global market, many countries continue to expand their grazing areas to compete.

The most well-known worst case is Brazil, whose cattle ranchers demolish rainforest acres to use as cattle grazing land. This practice not only increases livestock emissions but also destroys naturally occurring carbon capture ecosystems—a double whammy.

The type of beef produced in Brazil differs from American beef. America specializes in well-marbled, grain-finished beef, while Brazil exports grass-finished, leaner products. American beef is targeted to higher-income countries such as Japan, South Korea, and Taiwan, whereas Brazilian beef is marketed to lower-income countries like China, Chile, Egypt, and Iran, where demand growth is much faster.

Brazilian company JBS, the largest meat producer in the world, increased its annual greenhouse gas emissions by 51 percent between 2016 and 2021 (280 million to 421.6 million metric tonnes), based on the Institute for Agriculture and Trade Policy's latest 2022 calculations. JBS now produces more greenhouse gas emissions than all of the fossil fuel companies combined, or put another way, more than the country of Italy or France. It is substantial to say the least. JBS customers include global food chains like Burger King, McDonalds, and Walmart. In 2020, JBS pledged to make a 30 percent cut in greenhouse emissions within ten years, and in 2021 announced a goal of reaching net zero emissions by 2040; however, their actions show otherwise. "Some government and business leaders are saying one thing, but doing another. Simply put, they are lying, and the results will be catastrophic," remarked UN secretary-general António Guterres as he presented the 2023 report for the Intergovernmental Panel on Climate Change.[*] In response to the "net zero" pledge, the New York attorney general filed a lawsuit in March of 2024 alleging JBS continued to mislead consumers after trade associations NAD (National Advertising

---

[*] United Nations, "Secretary-General Warns of Climate Emergency, Calling Intergovernmental Panel's Report 'a File of Shame', While Saying Leaders 'Are Lying', Fuelling Flames," press release no. SG/SM/21228, April 4, 2022, press.un.org/en/2022/sgsm21228.doc.htm.

Division) and NARB (National Advertising Review Board) found a lack of a formulated net zero plan.

While the environmental toll of livestock ranching is becoming more socially recognized, the FAO still projects that global meat consumption will rise by more than 1 percent annually as low- and middle-income countries desire first-world tastes. Unfortunately, the majority of this meat will come from Brazil, at the expense of the rainforest. In turn, the FAO estimates that global emissions from food production will rise 60 percent by 2050!

It doesn't appear that synthetic meats will satisfy meat demand, and the resource footprint of these alternative meats is not yet known. Producing livestock with lower resource footprints is critical to solving the emissions problem, yet livestock is not the only agricultural challenge within our food production system.

Monoculture farming is a very modern method of agricultural production that consists of growing a single crop using the majority of the land (think giant rows of corn). The principal belief behind monoculture farming is that it is more efficient and profitable to provide the needs for just one species, as opposed to growing and harvesting multiple species. These profits come at the expense of the environment. Planting the same species year after year soaks the nutrients out of the ground, diminishes biodiversity, and allows specialized viruses to flourish.

One of the best examples of monocropping virus catastrophes is the southern corn leaf blight epidemic of 1970 and 1971, which destroyed 15 percent of the corn crops across America and Canada at a cost over $1 billion in 1970 (around $7 billion today). The mishap was almost one hundred years in the making.

Prior to 1930, virtually all corn produced in the world utilized open-pollinated cultivars, with about one thousand types of corn being grown. In 1909, two geneticists determined that inbreeding corn to establish pure lines and then crossbreeding those lines could lead to improved yields. Unfortunately, their parent stocks were low-producing varieties and therefore their experiment was a failure; however, the concept had merit. In 1926, Henry A. Wallace successfully bred high-yield corn varieties and founded the very successful company Pioneer Hi-Bred. Wallace eventually became secretary of agriculture and was later elected the thirty-third vice president of the United States. His term "hi-bred" quickly became nicknamed "hybrid" and is now the accepted name for any specific cross of plants or animals that results in an increase of heterosis of progeny (greater biomass, speed of development, and fertility than the parents).

By 1960, the amount of hybrid corn produced in the United States had increased to 90 percent. Almost the entire industry had placed their bets on the new genetic corn hybrids.

The first warning sign of a problem was reported in 1961 and confirmed in 1964 and 1965. Hybrid seed companies used areas of the Philippines and other tropical regions as winter nurseries. This allowed for continuous hybrid development and increases in seed stockpiles for sale to producers in the United States and elsewhere. This practice is still widely popular today. The exact source of the corn leaf blight inoculum that entered North America will never be known, but it likely entered through one of the tropical contaminated seed stocks.

By fall of 1969, two edges of the disease triangle, a virulent pathogen and a susceptible host, were in place. To trigger a major plant disease outbreak, all that was needed was a favorable environment for the leaf blight fungus. The third edge of the disease triangle fell into place in the winter of 1970. By the spring of 1971, species from Florida began to trickle into pathologists' labs with serious damage to leaf tissue. By August, the disease had engulfed most of the southern Corn Belt, as far west as the eastern halves of Nebraska and Kansas, up into southern Minnesota, and as far north as southern Ontario, Canada.

Estimated total loss was twenty-four million hectoliters of grain, and the total calories of food energy lost was greater than that lost due to the potato leaf blight in Ireland in the 1840s, which took as many as one million lives from hunger and disease and changed the social and cultural structure of Ireland.

Five decades have passed since the epidemic, and the catastrophe is rarely mentioned outside of plant pathology classes. Purdue University professor A. J. Ullstrup's warning in 1972 is as true today as it was then: "Never again should a major cultivated species be molded into such uniformity that it is so universally vulnerable to attack by a pathogen, an insect, or environmental stress. Diversity must be maintained in both the genetic and cytoplasmic constitution of all important crop species."[*]

With the recent developments in genetic engineering and the instability of climate change, one has to wonder if we are setting ourselves up for more disasters. Some estimates conclude that about 80 percent or 440 million acres of the United States is being cultivated for monoculture, much of it grown to feed livestock. Monoculture has already exacerbated problems with herbicide-resistant super-weeds and insecticide-resistant bugs. The mass-flowering of single plant species is increasing the prevalence of bee populations infected with parasites. Continued plantings of monoculture crops consume the same nutrients year after year, and therefore require large quantities of fertilizers and chemicals for sustained crop growth.

---

[*] A. J. Ullstrup, "The Effects of the Southern Corn Leaf Blight Epidemic of 1970–1971," *Annual Review of Phytopathology* 10 (September 1972): 37–50, doi.org/10.1146/annurev.py.10.090172.000345.

The overuse of chemical fertilizers destroys the ground crops, which typically help the soil to retain water and prevent erosion. Dry soil leads to increased erosion and the need for vast amounts of water to irrigate crops because the soil water retention is minimal. Increased irrigation leads to fertilizer runoff, which pollutes waterways with nutrients and causes species die-offs. Increased irrigation also leads to water shortages, and it takes just one drought year to throw an entire city like Cape Town into disarray. It's easy to see how the downward spiral can continue, city by city.

These agricultural problems are at the core of climate change, freshwater stress, pollution, ecosystem degradation, and wildlife loss. If we want to correct one or all of these problems, we *must* reinvent our food production systems. There is no other alternative to solve our global challenges and provide a habitable world for future generations.

IT TAKES ENERGY

THE MOST SUSTAINABLE AND PERMANENT source of energy in the solar system is our Sun, as it will provide about ten billion years' worth of energy—plenty enough for our species to learn to live sustainably on this planet and to move beyond to other planets. In order to create a sustainable society on Earth, we must first learn to harness the Sun in *all* of our production systems. It's not a far reach from our cavemen ancestors who had to learn how to harness fire.

In the past, humans decided to grow the plants and animals we loved the most, based solely on abundance, ease, use, and flavor, with zero forethought to the resource intensity of those organisms. Our ancestors foolishly believed the resource abundance of Earth would provide forever. Never did we think that we could overfish the oceans, strip the lands of nutrients, and change the weather with our actions. For nearly twelve thousand years, our agricultural evolution has not once taken into account the finite resources of the Earth. The widespread idea of resources being finite is truly only a decade or two old.

If we had known in the past that our resources were limited, we would have never invested in the resource-intense land animals and giant monoculture farms that we have today. If we had known that we would one day have to compete with our agricultural production systems for energy, land, and water, we would have never built the food system we have today. To solve the food challenges of the future, we must rethink the food production system from a resource consumption approach.

The comparison of resource usage among common foods is the first step in reorganizing our food production system. When looking to find a solution to our future food problems, we must identify species that are the most energy efficient and nutrient dense, release the least amount of greenhouse gases, and require the fewest number of finite resources for growth. The species that perform best in these criteria should form the basis of our food production system. As you'll soon discover, shellfish quickly become the standout winner.

By default, our food production system is already solar powered. Plants directly convert solar energy into vitamins, proteins, and minerals that meet our nutritional needs. Further up the food chain, animals eat the vegetation and create more compact and complex vitamins and proteins. Apex predators at the top of the food chain are the least efficient at nutritional solar conversion, relying on the animals beneath them to do the nutrient conversions for them. Using this measurement, it is easy to identify sources of nutrition that are more efficient than others at converting energy into nutrition. This scale of solar energy conversion into nutrition is the most fundamental and basic measurement of energy efficiency in food.

When we begin to account for the other sources of energy required, from start to finish, to bring our farmed food to the table, we quickly realize that energy efficiency should be a primary requirement for the foods of our future.

The United States consumed 100.41 quadrillion BTU (British thermal units) of energy in 2022, with the food system using about 10 percent of that total, about 10.11 quadrillion BTU. Put another way, the American food system consumes as much energy preparing and transporting food as it takes to power the country of France for an entire year.

Around the world, food systems consume about 30 percent of the total energy consumption, primarily using fossil fuels, which in turn account for about 20 percent of the global greenhouse gas emissions.

The four parts of energy use in our food production are: **handling** (49.5 percent), **agriculture** (20.8 percent), **processing** (15.8 percent), and **transportation** (13.9 percent).

Food handling is the largest consumer of energy, including packaging, retail, restaurants, and consumers. Examples include milk and egg cartons, or cans and wrappers, and the refrigeration used to keep perishable products stable in restaurants and homes. Shellfish register rather low in this category as they are predominately eaten raw in the shell, which means no packaging needed. Mesh bags or boxes are used as containers to deliver oysters, with many of these being compostable or reused by the farmer. Being a perishable product, they do require refrigeration, but also do not travel far distances from the farm and have a short shelf life, typically consumed within days of delivery. A high-volume restaurant is receiving shipments of the same oyster multiple times a week to ensure fresh product. In very high-volume situations, our farm delivers daily to the restaurant, reducing the need for prolonged storage and energy use.

Agriculture includes activities in the growth and cultivation of food. Sixty percent of this energy is consumed directly in fuel and electricity use, while the rest is indirect through fertilizer and pesticide production. Shellfish require zero fertilizer and pesticides, while also demanding very little energy and fuel consumption for growth. The only power consumption on a shellfish farm is the nursery system, which may use a small pump. Our farm utilizes tidal power for the nursery, and therefore requires zero energy usage.

Food processing is the transformation of raw ingredients into food products, such as wheat into flour. Unless sold as a shucked meat product, shellfish do not require processing. When they are shucked, it's by manpower, not electricity.

Transportation from farm to table is the lowest energy use. Transportation incorporates any energy used by delivery vehicles moving product from the farm to the end user. For a typical shellfish operation, a delivery van and a small 4-stroke outboard boat engine are the only vehicles to use fuel. Depending how far the oysters travel, additional trucks may be used for transport.

One of the most interesting aspects of looking at food production and the resources required to produce that food, is that they all differ greatly. Not all foods are created equal. As a general rule, raw foods require the least amount of processing and therefore require less energy than processed foods. Raw foods are better for you than processed foods, and they are also better for the planet.

Some of the most energy-efficient foods include wheat, beans, fish, eggs, nuts, and shellfish. The most resource-intense foods are typically animal-based products. Beef, being the worst, requires up to twenty times more resources and emits twenty times more greenhouse gases than plant-based protein sources. Even some plants such as roots and tubers require disastrous amounts of fresh water and land-use change. One of the solutions to our future food challenges is to replace resource-intense products with more efficient ones.

With little surprise, ocean organisms are some of the least-re-source-intense food products on Earth. At present, seafood is considered one of the most carbon-efficient foods on the planet, with wild-caught fish requiring no land, fresh water, or feed inputs, while having little to no impact on wildlife. Obviously, some methods of wild fishing are better than others, and some countries have better stock management strategies than others, so it's important to refer to the third-party organizations who monitor these issues. In general, seafood is a smart choice when it comes to resource-use efficiency.

Of all seafood products, shellfish are by far the most sustainable, requiring zero fresh water, zero land, and zero fertilizers. Shellfish products are typically eaten raw or cooked, with minimal to no processing required. The shells of shellfish products even sequester carbon, making them one of the only food products capable of having a neutral or negative carbon footprint.

Farmed shellfish can literally feed the entire world. Around 40 percent of the U.S. population lives along the coastline, providing direct access to ocean farm products, and within twenty-four hours these products are shipped to landlocked locations via refrigerated transport. Professor Ronald Osinga at Wageningen University in the Netherlands believes that a global network of sea-vegetable farms totaling 180,000 square kilometers (about the size of Washington State) could provide enough protein for the entire world! For perspective, current worldwide farmland consumes an area roughly the size of Canada and the United States put together and *still* does not meet the protein needs of the world.

If we want to transform our food production system to grow the products that require the fewest number of resources, shellfish farming would be at the top of the list. In addition to feeding the human population, shellfish can also become feed and fertilizers for our existing food structure. Shellfish can help restore our agricultural soils, reduce our need for synthetic (high energy use) fertilizer and chemical inputs, lower the resource footprints of existing livestock, combat ocean acidification, and sequester carbon. This is all possible without the need for land, fresh water, or energy inputs.

SOLUTIONS

I
F THE WORLD WANTS TO BE PROACTIVE IN ESTAB-
lishing food productivity to meet our growing demands, the
most logical solution is to utilize aquaculture and our oceans.
This abundance of untapped resources and space has the ability to
not only meet our future needs but also transform our existing food
production system into a more efficient and productive practice.

In their 2022 *State of World Fisheries* report, the FAO makes
the argument multiple times that

> *In the next ten years, aquaculture must expand sustainably to
> satisfy the gap in global demand for aquatic foods, especially
> in food-deficit regions, while generating new or securing exist-
> ing sources of income and employment. This requires updating
> aquaculture governance by fostering improved planning, legal
> and institutional frameworks and policies. FAO and its part-
> ners must focus on the urgent demand for the development and
> transfer of innovative technologies and best practices to gener-*

ate efficient, resilient and sustainable operations. The contin-
ued transformation of aquaculture applies to most regions but
is particularly critical in food-insecure regions; the aim is to
increase global production by between 35 percent and 40 per-
cent by 2030, according to national and regional contexts.[*]

To date, much of the aquaculture sector has been left to private enterprise to develop and explore. Most shellfish farms are small family farms that supply their local shellfish demand and are unable to scale to meet inland demand. A lot of farms cannot even get started due to cumbersome permitting and zoning requirements. To aid these startup farms, governments should make zoning and permitting of aquaculture resources less burdensome—like land farming.

If we are going to scale up the industry as quickly as possible (which needs to happen), we need to streamline the permitting process. Individual towns need to take the first steps in mapping appropriate sites around their coastal waters that would achieve the greatest success rates. Municipalities can eliminate the burden of forcing individuals to guess which sites will be the most productive, or whether the site will be approved by the town and neighbors, by preapproving sites. Townships ultimately have the final authority on the zoning of aquaculture sites, so why not pre-zone areas to help speed up aquaculture economic development? Every forward-thinking town should zone their appropriate waters, creating preapproved farm zones and regulations in anticipation of an applicant.

The implementation of preapproved zones will not only promote aquaculture applications and opportunities to town residents, but it would also eliminate a lot of the controversy and

---

[*] FAO, *The State of World Fisheries and Aquaculture 2022* (Rome: FAO, 2022).

NIMBYism that comes with most applications. Applicants and neighbors will know exactly where preapproved areas are zoned, and which type of farm gear will be permitted. This simple act can expedite the process by years.

Every coastal state has a fisheries management organization, which is typically responsible for zoning the initial water quality maps. These agencies decide which "grade" of water quality a specific watershed receives, and whether harvesting or farming of shellfish or other species is allowed or prohibited. These agencies hold the fishery knowledge. It would be very simple for these agencies to have an extension officer who, at the request of a town, could take a closer examination of the municipality watersheds and suggest prospective farm zones. At a minimum, each town could start by only considering the "approved" water quality areas for pre-approved aquaculture sites.

Each coastal town has a committee (sometimes multiple committees) that oversee watershed construction (like docks) or fisheries management. These local agencies should be the leaders in taking state fishery suggestions and implementing them on the local level. Coastal towns need to zone their waters in anticipation of aquaculture farms, and they need to promote these opportunities to residents. Shellfish aquaculture will not only boost the local economy by creating jobs and a new food source, but it will revitalize the local environment and improve water quality. The towns in my area are currently taking this approach in an attempt to support and scale our local industries. Within the next year we should have preapproved aquaculture zones in our surrounding waters.

Government solutions tend to be the slowest solutions. Luckily, the Nature Conservancy and the Pew Charitable Trusts have partnered together to implement a program to identify productive areas within coastal watersheds ripe for the renourishment of oyster reefs, and have begun to restock these areas using oysters from aquaculture farms around the nation.

Started during the Covid pandemic as a way to supplement oyster farmer incomes while markets were closed, the Supporting Oyster Aquaculture and Restoration (SOAR) program purchases oversized oysters from farmers and plants them in ideal locations to restore coastal ecosystems.

In the first two years of the project, SOAR redirected 3.5 million oysters from farmers to twenty-five sites across the country, encompassing forty acres of oyster reefs while supporting 125 shellfish companies. These restored reefs will armor the shorelines against coastal storms and erosion, remove nitrogen to improve water quality, provide habitat for juvenile marine species, sequester carbon within the oyster shells for centuries, and provide sustainable food for humans and animals alike. All of these benefits come without the need of land, fertilizer, or fresh water. Everything wins.

On my farm, we have partnered with the Woods Hole Oceanographic Institution and Southern Connecticut State University to conduct ongoing research into how our subtidal bottom-cage farm interacts with the environment. The project is showing very promising data that confirms our suspicions: subtidal shellfish aquaculture farms not only provide adequate nursery areas for wild marine species, but also enhance and help propagate these wild resources! The farm is literally rebounding the marine ecosystem.

While controversial, municipalities should create "no fish zones," protected marine areas in which oyster reefs can be reestablished and where marine species can thrive. Like land conservation parks, we need to give our marine wildlife areas where they are protected. Infusing these areas with protected oyster reefs will aid the restoration of species and the ecosystem.

In proof of concept, our farm research project has shown an enormous amount of accumulation of biodiversity around our shellfish cages. These structures create habitat for juvenile fish species, aiding in the rebuilding of depleted wild fish stocks. It's no coincidence that fishermen surround our farm with lobster, fish, and whelk traps. Our oyster farm acts as a nursery sanctuary for these wild species by providing food and shelter. These wild species use the farm as a home, and the fishermen are then able to capture them once they've ventured off the farm and reached market size.

In their 2020 article, "Sustainable Bivalve Farming Can Deliver Food Security in the Tropics," professors of zoology at the University of Cambridge Dr. David Aldridge and Dr. David Willer state:

> *Bivalve reefs (and bivalve farms, during the period between harvests) can buffer estuaries and coastal waters against phytoplankton blooms caused by anthropogenic nitrogen loading, increase water clarity, provide a nursery habitat for fish, provide coastal flood and storm protection, and shell production acts as a form of carbon capture. The ecosystem services yielded from bivalve aquaculture are currently estimated at US $30.5 billion per year and only set to grow as the industry expands.* [*]

Shellfish are beginning to be viewed as a solution to our declining marine ecosystems.

Another groundbreaking shellfish activity is taking place on

---

[*] David F. Willer and David C. Aldridge, "Sustainable Bivalve Farming Can Deliver Food Security in the Tropics," *Nature Food* 1 (July 2020): 384–388.

Martha's Vineyard, which if replicated around the world could help speed the development of shellfish culture. Each town on the Vineyard pays $30,000 to a small shellfish hatchery annually, in exchange for several hundreds of thousands of baby oysters, clams, and bay scallops. These babies are grown in a town-owned nursery for the summer and then released into our ponds and estuaries, where they will overwinter and spend the next year ripening to market size. This resource not only provides townspeople with a source of recreation and food gathering, but the shellfish also provide for cleaner waters and habitats for juvenile marine organisms.

Every coastal town in the world should belong to a program like this. First and foremost, our local governments should invest in the production of sustainable, highly nutritious foods that better our environments. The use of tax dollars to support and purchase shellfish from a local hatchery promotes and safeguards local jobs within the industry while improving the environment and providing food. There is no better way to use taxes. It's a no-brainer. Get your town on board!

The Billion Oyster Project, an organization that has restored oysters at over fifteen reefs in the five boroughs of New York, views the oyster as a natural solution to rising oceans. With over a decade of work, the Billion Oyster Project is beginning to make long-standing impacts. Over two million pounds of oyster shells have been collected from local restaurants by the organization, which recycles them to create the foundation of new and restored oyster beds. The beds were seeded with the aid of Fishers Island Oysters and are now thriving in and around the New York harbor. The oysters have recently begun spawning and recruiting natural seed sets on their own. A visit in September of 2023 by the Prince and Princess of Wales brought global attention to the project, with the aim of starting similar projects in all major cities around the world.

In Louisiana, over six miles of oyster reef has been deployed by the Nature Conservancy to restore wildlife habitat, protect shorelines, and rebuild the wild oyster fishery. A similar project was successful in the Netherlands.

In Bangladesh, the island of Kutubdia has seen the greatest success. Rings of concrete covered in oysters have been placed in coastal regions that have witnessed coastal erosion at rates surpassing most global locations. It is estimated that by 2050, one in seven Bangladeshis will be displaced by climate change; however, the oyster reef rings have shown promise. Up to twelve inches of sediment accumulation have been documented behind the reef project, and commercial fish species, which haven't been present in years, have suddenly shown up within the artificial oyster reef. Efforts are underway to expand the project.

In Maryland, oyster farms are being included in a water quality trading program, where "dirty" companies that create excess nutrients can buy "clean" credits from oyster farms to help reduce their environmental footprints. Research from the Virginia Institute of Marine Science has found that an acre of restored oyster reef can remove about five hundred pounds of nitrogen each year. Furthermore, 1,300 acres of oyster reef would equal the same effects as a new wastewater treatment plant.

The town of Mashpee in Massachusetts has taken a similar approach, using the harvest of five hundred thousand oysters a year in their nutrient management plan to remove about 2.5 metric tons of nitrogen. In Denmark, mussels are used for the same eutrophication mitigation (nutrient removal) strategy.

Using shellfish as a solution to improve our local marine environments seems to be catching on and the data is there to support it. Governments around the world are now backing shellfish programs for environmental restoration projects, and it makes sense;

restoring the pillar species of the ecosystem that we have harvested to near extinction will only serve to help rebalance and restore those degraded ecosystems. The job moving forward is to not only repair these degraded ecosystems, but to also learn how to exploit them (for food) in sustainable ways.

The continued expansion of the shellfish farming industry and the restoration of wild oyster reefs will undoubtably provide sustainable food for our growing population and increase local food production jobs while cleaning our watersheds and making them healthier and more attractive for a number of marine species.

Only two obstacles stand in the way of this expansion: water quality and demand.

If water quality around the world degrades, so too will the viability of ocean farms. Like their land counterparts, ocean farming depends on a clean environment. It's imperative that nations continue to protect and maintain clean ocean environments.

In 2022, the United Nations' Intergovernmental Oceanographic Commission outlined the current threats facing our oceans. The highlights include climate change (ocean acidification and sea water levels rising), plastic and ocean pollution (such as land runoff, oil spills, and ocean dumping), fishing and fishing gear, shipping and transport, offshore drilling, and deep-sea mining.

While the list seems demoralizing, the Commission also outlined a series of solutions to these problems, such as establishing additional marine reserves and parks, reducing and preventing pollution from entering the oceans, eating less fish and meat, converting industries to more sustainable energy types, and lowering transport emissions.

Impressively, the UN has "walked the walk" on the forefront of ocean stewardship. In 2008, 191 nations within the UN agreed to a moratorium on large-scale commercial fertilizer schemes to mitigate climate change. This agreement called for a ban on major ocean fertilizer projects until scientists could better understand the risks involved.

In June of 2023, 193 nations passed an agreement on the Law of the Sea, relating to the conservation and sustainable use of marine biological diversity in areas beyond nation jurisdiction—basically, an agreement governing the high seas in which no nation has jurisdiction. This is the first agreement to ensure the Earth's oceans are protected for generations to come and comes with roughly $857 million from the EU to enable success. A critically important aspect of the treaty is that it creates international marine protected areas where destructive activities could be restricted. In these protected areas, polluters and rule breakers would end up paying penalties, which could serve as a means to the end of certain practices that contribute to climate change, biodiversity loss, and pollution.

The UN agreement comes on the heels of the $1 billion Biden Administration's 2021 "America the Beautiful" executive order to restore, connect, and conserve 30 percent of lands and waters by 2030. Biden's program looks to build upon the National Parks initiative, expanding it to also include waterways. Taking a step beyond borders, Biden called on other nations around the world to meet the 30x30 challenge.

The math is simple; if the ocean dies, so will humanity. There just isn't enough land and resources to sustain our population. This notion is finally being recognized, as attested by the 193 nations signing onto the UN treaty. However, keeping our oceans clean is not enough. Demand for shellfish products needs to continue to increase for decades to come. The more shellfish we can put in our environments, the healthier these environments will become, and the closer we can come to meeting our global food nutritional needs.

We have zero time to waste. The potential for aquaculture will most likely decline over time due to climate change temperature changes and the resulting decrease in phytoplankton content. Ocean acidification threatens to dissolve larvae shellfish before they can reach maturity. These effects will be slowed by implementing shellfish farms around the world as quickly as possible. When shorelines are lost to rising seas, these areas should quickly be converted to new shellfish farms to help mitigate further coastal erosion and to solidify food security for the region. The sooner we can expand the industry, the better.

To be the most sustainable protein farming practice on Earth, after the shellfish meats have been consumed, the shells themselves need to be utilized in as many products and industries as possible. This will help the industry expand even faster. Human consumption for food is just the tip of the iceberg. I'm talking about using the shells for building materials, clothing, beverages, furniture, animal feeds, fertilizers, and more.

Below I highlight unique companies, projects, and innovations that are elevating shellfish into the future. We can learn from these examples, support them, duplicate them, and expand on them. The more we can expand the shellfish industry and incorporate shellfish into existing industries, the better off the planet and humanity will be. The path toward global stability begins with shellfish farming.

ANIMAL FEEDS

THE OCEAN IS NOT AN INFINITE SOURCE OF food. Around 85 percent of global fish stocks are over-exploited, fully exploited, depleted, or in recovery from exploitation. All seventeen of the world's major fishing areas have reached or exceeded their natural limits.

Over one-third of the world's fish catch is fed directly to livestock such as pigs, chickens, and farm-raised fish. This is yet another example of our traditional food system competing with our direct needs. By substituting shellfish into these animal feeds, we can divert a portion of our wild fish harvest directly to human nutritional needs and allow the wild stocks to replenish.

When used as a feed, fish are typically incorporated into a product called fishmeal. Fishmeal is simply the process of drying fish and crushing it into a powder or cake. The practice of feeding fishmeal to livestock is ancient. A primitive form of fishmeal was described by Marco Polo during the fourteenth century, when dried fish was fed to cattle, sheep, camels, and horses.

Currently around six million tons of fish are harvested annually to make fishmeal. Common fish species targeted for fishmeal include anchovies, sardines, menhaden, pout, mackerel, and sand eels. Peru produces almost one-third of the total fishmeal supply. Unfortunately, it takes four to five tons of fish to produce just one ton of fishmeal.

Typically made from human-consumable fish or commercial fishing bycatch, large-scale fishmeal production is controversial in that it encourages corporate fisheries not to limit their yields of bycatch and to instead try to target bycatch on purpose. These practices have led to the overexploitation of ocean resources and the current state of collapsed fisheries we see today. Most fishmeal species form the bottom of the food chain, and through overharvest of this food supply, we unintentionally destroy all of the other species that depend on them. These fishing practices are one of the reasons the ocean is in the mess that it is.

We do not have to give up our fish, chicken, pork, or beef; we just need to switch the feed of these animals to a shellfish-based fishmeal diet. We can use the environmental benefits of shellfish to reduce the footprints of traditional farmed products by incorporating them into our traditional animal feed systems.

For example, chickens who consume shellfish in their diets have a lower environmental footprint when compared to corn-fed chickens because the shellfish feed requires fewer inputs to produce than the traditional corn feed. This same "trickle up" effect can be applied to all fishmeal-fed farm operations. In essence, any animal feed that requires protein could be supplemented with shellfish to reduce the environmental cost of the feed.

One such product has been developed in the Philippines and is currently available year-round. "Oyster powder," which is basically dried oyster meat in powder form, has been tested to replace fishmeal in Nile tilapia diets. At the conclusion of the eight-week trial, it was found that oyster powder could replace as much as 63.8 percent of dietary fishmeal by weight and would deliver the same growth results in the farmed tilapia fish.

It's widely known that chickens are fed a supplement of oyster shells to help make their eggs strong, but what if they were fed the meats as well? In New Zealand, a recent experiment giving mussels as feed to chickens has yielded surprising results. Not only does it lower the chickens' resource footprint, but the chickens also produce a healthier, more nutritious egg product.

New Zealand free-range egg producer Frenz includes green-lipped mussels in their chicken feed and by doing so has increased the omega-3 content significantly in their chickens' eggs. Levels of omega-3 "good fats" EPA and DHA were 77 milligrams per 100 grams in their standard chickens, while the shellfish-fed chickens produced eggs with 162 milligrams per 100 grams. Over double the amount of healthy omega-3s! At present they use 5 to 8 percent mussel feed in the chickens' diet and continue to scale up the amount of shellfish while recording results. I personally have fed mussels and oysters to my chickens and can attest to the awesomeness of the eggs. The yolks become as orange and rich as can be.

Even algae and seaweeds can help reduce the environmental footprint in animal feeds. Kelp meal, a product made of dried kelp seaweed, has been shown to help build beneficial antimicrobial and anti-stress properties in beef cattle, improve weight gain in Jersey calves, and increase milk production and omega-3 fats in Holstein cows.*

---

* Nicole Antaya and André F. Brito, "Results from Short-Term Studies Using Kelp Meal as a Supplement to Dairy Animals at the University of New Hampshire," NODPA, 2018, nodpa.com/n/244/Results-from-Short-Term-Studies-Using-Kelp-Meal-as-a-Supplement-to-Dairy-Animals-at-the-University-of-New-Hampshire.

Furthermore, a single cow expels about 220 pounds of methane each year, the greenhouse gas equivalent of burning nine hundred gallons of gasoline. Multiple studies have now shown that feeding dairy cows seaweed can reduce their methane outputs from a range of 50 to 90 percent! Such results have prompted states like California to look into using seaweed in dairy cow feed to help meet their 2030 climate goals. The addition of the seaweed into the cow's diet also allows for the elimination of some of the traditional resource-intense feed inputs.

Demand for sustainable feed products is growing. A 2023 market report estimated the global oyster powder market to be valued at about $934 million; however, this was expected to grow to over $2 billion by 2033, exhibiting a compound annual growth rate of 8 percent over the next decade.

The trajectory for shellfish feed to replace a large portion of our traditional fishmeal feed is slowly being set in motion. For shellfish feeds to succeed, it will take either continued consumer demand, or the collapse of the fishmeal fishery industry. We have the ability *now* to implement these shellfish feeds into the industry *before* the fishmeal fisheries collapse. Allowing them to collapse is foolish and will put us on a worse path forward.

Either way, shellfish will consist of a large portion of our animal feed and will provide positive environmental and health benefits up the entire chain. The strategic goal for our food system moving forward should be to implement these shellfish and seaweed feeds into as many existing niches as quickly as possible.

EAT YOUR
CARBON FOOTPRINT

NOT ONLY CAN WE SUPPLEMENT LAND-BASED agriculture feeds with shellfish, but we can also reduce our own resource footprints by implementing more of these species into our diets.

One of the greatest impacts that you can have on the planet is the food you choose to eat. The vast majority of planetary resources are used to produce, transport, and store food. By making smarter food choices, we can reduce carbon emissions, repair ecosystems, and boost local economies.

The Paris Agreement's central goal is to limit global warming to 2 degrees Celsius or less. However, the 2020 paper entitled "Global Food System Emissions Could Preclude Achieving the 1.5° and 2°C Climate Change Targets" states:

Even if fossil fuel emissions were immediately halted, current trends in global food systems would prevent the achievement of the 1.5°C target and, by the end of the century, threaten the achievement of the 2°C target. Meeting the 1.5°C target requires rapid and ambitious changes to food systems as well as to all nonfood sectors. The 2°C target could be achieved with less-ambitious changes to food systems, but only if fossil fuel and other nonfood emissions are eliminated soon.[*]

We have to change our food production systems. Food production generates about 35 percent of total global manmade greenhouse gas emissions. Of the 35 percent total, animal-based foods like meat, poultry, and dairy products, including crops to feed livestock, contribute 57 percent. Plant-based foods for human consumption contribute 29 percent while the other 14 percent comes from non-food products such as cotton and rubber.

Annually, the production of synthetic fertilizers and pesticides contribute more than one trillion pounds of greenhouse gas emissions into the atmosphere. The majority of the nitrogen from these fertilizers ends up in our oceans, with nitrogen levels now 50 percent above normal levels.

A recent World Bank study found that dedicating just 4.4 percent of U.S. oceans to only seaweed farming would absorb 10 million tons of nitrogen, 135 million tons of dissolved carbon, and 1 million tons of phosphorus, while providing 50 million tons of protein and sparing roughly 14 percent of freshwater withdrawals

---

[*] Michael A. Clark, Nina GG Domingo, Kimberly Colgan, Sumil K. Thakrar, David Tilman, John Lynch, Inês L. Azevedo, and Jason D. Hill, "Global Food System Emissions Could Preclude Achieving the 1.5° and 2°C Climate Change Targets" *Science* 370, no. 6517 (2020): 705–708.

and 6 percent of global cropland.* Substituting shellfish and sea-weed for synthetic fertilizers can also create major positive impacts.

According to the FAO, the livestock industry is responsible for 18 percent of global greenhouse gas emissions. Five million acres of rainforest are felled every year in South and Central America alone to create cattle pasture. Roughly 20 percent of all currently threatened and endangered species in the U.S. are harmed by livestock grazing. America's livestock and poultry farms produce about thirteen times more waste than the entire U.S. human population.† It's no wonder that animal agriculture is a chief contributor to water pollution.

The most resource-intensive feed animal currently in farm pro-duction is the cow. Beef is the largest food contributor to climate change, generating 25 percent of total food emissions, followed by cow's milk (8 percent) and pork (7 percent). It takes roughly 2,500 gallons of water, twelve pounds of grain, thirty-five pounds of top-soil, and the energy equivalent of one gallon of gasoline to produce one pound of feedlot beef. One pound! It is estimated that fifty bil-lion one-pound burgers are consumed a year by Americans alone. You do the math. It adds up to an enormous use of resources.

Plant-based foods are much better than animal products, yet they can still carry a heavy resource footprint. Rice is the largest crop contributor to climate change, responsible for 12 percent of global methane emissions and 1.5 percent of total greenhouse gas emissions. Following rice is wheat (5 percent) and sugarcane (2 percent). Rice takes the top seat due to the practice of farmers flooding their rice fields to kill weeds. These flood waters create

------

* "Seaweed Aquaculture for Food Security, Income Generation and Environ-mental Health in Tropical Developing Countries," World Bank Group, 2016.
† "What Happens to Animal Waste?" FoodPrint, October 8, 2018, updated February 28, 2024, foodprint.org/issues/what-happens-to-animal-waste.

ideal conditions for certain bacteria that emit methane gas. After rice is harvested, stalks and debris are left behind and are usually burned by the farmer to clear the fields for a new planting. Burning of the fields emits carbon dioxide, methane, carbon monoxide, nitrogen oxides, and sulfur oxides.

In comparison, ocean-farmed products use just a fraction of the resources of their land counterparts. In particular, farmed shellfish use the lowest number of resources. Not only are their resource footprints small, but shellfish crops also help combat ocean acidification and climate change as they grow by sequestering carbon and stabilizing ocean pH. Shellfish farming produces protein and vitamins for human consumption, without competing for human resources, while also bettering the ecosystems and environments they are grown in. It gets no better than that.

Measure Your Foodprint

NUTRITICS.COM AND THEIR "foodprint" carbon calculator can be used by chefs to understand in real time the carbon footprint of the meals being prepared. This service was most notably used for the COP28 event held in the United Arab Emirates. Foodprint data (printed on a menu) could help conscious consumers make better food choices in the future. For instance, Americans consume about sixty pounds of beef a year. A 2019 study found that if Americans replaced just 10 percent of their beef consumption with oysters, we could remove the equivalent of 10.8 million cars worth of greenhouse gas emissions.[*]

For the top five most sustainable foods in the world, Nutritics

---

[*] Nicholas E. Ray, Timothy J. Maguire, Alia N. Al-Haj, Maria C. Henning, and Robinson W. Fulweiler, "Low Greenhouse Gas Emissions from Oyster Aquaculture," *Environmental Science Technology* 53, no. 15 (2019): 9118–9127.

currently ranks them as: mushrooms, legumes, mussels and sustainably farmed shellfish, seaweed, cereal and grains, and organic fruit and vegetables.

But even with the rich global history of oyster and shellfish consumption, looking at modern-day statistics would have you believe shellfish have never been a significant source of protein. One of the biggest challenges for the bivalve shellfish farming industry is just to get recognized!

For example, the United Nations website boasts this chart, which illustrates various types of food and the associated greenhouse gas emissions it takes to produce them.

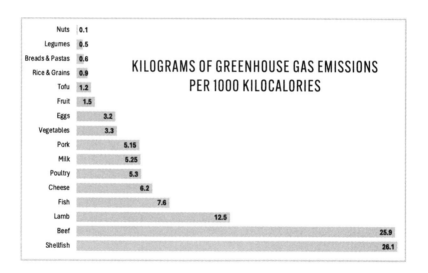

Source: United Nations

With all we know about bivalve shellfish farming, how could shellfish be labeled the worst?

You would have to read the fine print to discover that in this case, "shellfish" refers to shrimp farming.

According to the UN website: "Shrimp farms often occupy

coastal lands formerly covered in mangrove forests which absorb huge amounts of carbon. The large carbon footprint of shrimp or prawns is mainly due to the stored carbon that is released into the atmosphere when mangroves are cut down to create shrimp farms."*

Without proper labeling such as "shrimp farms" or "bivalve farms," most people see a chart like the UN example and instantly associate oyster farming as a harmful farm practice. The industry won't grow fast enough with this kind of mixed messaging. Unfortunately, almost all modern food industry emission data comparisons fail to include bivalves in their data points. To help bolster the industry, we need bivalve proteins to be specifically included on these lists. (FYI, consumers should use "eggs" as the most similar data set to bivalve farming when looking at these charts.) The shellfish industry needs a proper lobby, and an "optics" team to help manage this type of oversight.

On the nutritional side, bivalves (oysters, mussels, scallops, and clams) have a higher protein content than beef, are rich in omega-3 fatty acids, and have some of the highest levels of minerals of all animal foods. A dozen oysters contain about twenty-four grams of protein, equivalent to a quarter-pound hamburger. Just six oysters a day meet the daily intake requirement of iron, zinc, copper, iodine, magnesium, calcium, and phosphorus. Oysters are natural multivitamins for humans.

On a scale of micronutrient density, oysters rank second only behind beef liver. A one-hundred-gram serving of oysters (about eleven oysters) also contain the following Reference Daily Intake (RDI) values:

---

* United Nations, "Food and Climate Change: Healthy Diets for a Healthier Planet," un.org/en/climatechange/science/climate-issues/food.

Vitamin D: 80% of the RDI
Zinc: 605% of the RDI
Thiamine (vitamin B1): 7% of the RDI
Niacin (vitamin B3): 7% of the RDI
Vitamin B12: 324% of the RDI
Iron: 37% of the RDI
Magnesium: 12% of the RDI
Phosphorus: 14% of the RDI
Copper: 223% of the RDI
Manganese: 18% of the RDI
Selenium: 91% of the RDI

Just 68 calories worth of oysters (about six oysters) give you a week's worth of immune-boosting zinc and one to three days' worth of vitamin D, selenium, copper, and vitamin B12! Oysters are rich in all three types of omega-3s (ALA, DHA, and EPA) and contain a recently discovered antioxidant called DHMBA, which is fifteen times more effective in reducing oxidative stress than vitamin E.

Cambridge University professors Willer and Aldridge claim that "if just 1 percent of the area in the tropical regions alone would be developed, over 120 Mt [metric tons] of bivalve meat could be produced annually—enough to satisfy the protein demands of approximately 715 million people."[†]

Shellfish are extremely beneficial to our health. In comparison, red and processed meat was responsible for $285 billion in health-care-related costs in 2020. Three-quarters of this figure is caused

---

[*] Michael Joseph. "9 Health Benefits of Oysters (and Full Nutrition Facts)," Nutrition Advance, September 26, 2023, nutritionadvance.com/oysters-nutrition-benefits/.

[†] Willer and Aldridge, "Sustainable Bivalve Farming."

by the consumption of processed meats (bacon, sausage, salami, ham, beef jerky, and canned meat, for example). Processed meats have been linked to a myriad of diseases, including colorectal cancer, rectal cancer, breast cancer, liver disease, cardiovascular disease, diabetes, and other health issues like manic episodes.[*]

One of the more groundbreaking and interesting studies on oysters today involves looking closely at the oyster immune system and the microorganisms that flourish within them. How is it possible that an animal that has lived hundreds of millions of years without the need to evolve has been able to overcome a plethora of ever-evolving viruses, bacteria, climates, and enemies?

New research suggests that oysters receive a "default" immune system from their parents, but they are able to also adjust and adapt a new immune system throughout their lives by interacting with the microorganisms present in their environments. As the surrounding animals and environments evolve, so too does the oyster's microbiome and immune system. These changes are passed on to the oyster offspring, who start off with the parents' default immunity, and then continue to evolve new immunity with the changing environment. When we ingest oysters, we too absorb this microbiome, helping to boost our own immunities and microbiome in turn.

Eating shellfish is clearly a healthy choice that will reduce the financial burdens on the healthcare system, improve the environment, and combat climate change. Replacing just a handful of your protein meals a year with a shellfish product could have enormous positive implications for the future.

---

[*] Marco Springmann, Daniel Mason-D'Croz, Sherman Robinson, Keith Wiebe, H. Charles J. Godfray, Mike Rayner, and Peter Scarborough, "Health-Motivated Taxes on Red and Processed Meat: A Modelling Study on Optimal Tax Levels and Associated Health Impacts," *PLOS One* 13, no. 11 (November 2018).

To help cut down on worldwide resource use, try limiting your beef meals to just once a month, or try implementing the top six most sustainable foods into your monthly diet. Involve your children when searching for and cooking new recipes. It's one of the smartest activities we've done as parents. Soon you'll find the kitchen to be the best classroom available, and dinners together become fun activities and lasting memories. Talk about the food you eat and where it comes from. Try to reconnect with your foods' origin, and you'll open the door to an amazing and worthwhile lifetime adventure.

Instead of ordering chicken wings for an appetizer, choose oysters, mussels, clam cakes, or calamari.

Instead of chicken noodle soup, choose clam chowder.

Instead of a hamburger, choose an oyster po'boy, a mushroom burger, or a fish sandwich.

Instead of meatballs in your pasta, choose mussels or clams.

It's easy! When you start to think about your food intake, it quickly becomes apparent that some choices are better than others. Your body will thank you, your planet will thank you, and most importantly, your grandchildren will thank you.

CORPORATE RESPONSIBILITY

OUR SPENDING CAN EMPOWER MOVEMENTS and sustainability. Who we support with our money matters. One of the best ways we can influence positive change is to support the companies and products that inspire the positive change we hope to see. The following companies are industry leaders who support the shellfish industry in unique and inspiring ways. Support them!

WINE

IF YOU WANT to support impactful nonprofits while having a glass of wine with your clam pasta, look no further than wine company Proud Pour (proudpour.com), which donates 5 percent of their top-line revenue to a slew of nonprofits around the nation making positive change.

From oyster reef, bee habitat, and coral reef restorations, to providing food for injured sea turtles and supporting restorative farming practices, Proud Pour has created a model that *every* business could and should follow.

I've had the chance to meet and work with these folks, as one of the organizations they happen to support is the Martha's Vineyard Shellfish Group, our island nonprofit organization that is responsible for rebuilding wild oyster reefs and shellfish populations within our salt ponds. They are an amazing group of people, and in addition to a great give-back program, they also have incredible wine! Order some up.

FASHION

MAYBE YOU LIKE to express yourself through fashion. Clothing company Jetty (jettylife.com) has developed a material made from pulverized oyster shell power and recycled plastic bottles, which are combined into pellets, then melted and strung into fibers. The resulting "Oystex" fabric is spun into shirts, shorts, jackets, sweaters, and more.

In addition to incorporating oysters into garments, the company started a nonprofit organization, the Jetty Rock Foundation, to donate funds and energy toward rebuilding New Jersey's oyster reefs.

Another brand utilizing recycled oyster shells and water bottles in their garments is the Long Wharf Supply Co. (longwharfsupply .com), which was featured on *Shark Tank* for their innovative fishing sweaters. Each sweater reseeds up to thirty oysters.

How about a pair of sunglasses made of out recycled oyster shells? Made by French company Friendly Frenchy (friendlyfrenchy .fr), the "shell sunglasses" boast a handmade wood cellulose and recycled shell frame.

COSMETICS

ONE OF THE newest inroads for oysters is the cosmetic industry, yet with all the vitamins and nutrients it doesn't seem surprising. When used in cosmetics, the shell tends to be the most utilized ingredient, replacing microplastic beads with ground-up shell powder for exfoliating scrubs.

Brands such as Aquatonale Group (aquatonale.com), PhycoHealth (phycohealth.com), Perlucine (perlucine.fr), Thermes Marins de Saint-Malo (cosmetique-thermesmarins.com), and Ostrealia (ostrealia.fr), as well as individuals on Etsy, all sell exfoliating scrubs utilizing oyster shells. These are all great replacements for microplastic beads.

Some of the more unique cosmetics come from brands such as Edulis (edulis-cosmetics.com), which has created oyster oil extracts for anti-aging creams, balms, serums, and gels.

SPREAD THE CULTURE

WE SHOULD ALL STRIVE TO MAKE POSITIVE changes in the world, but it seems the hardest part is knowing what we can do. One of the easiest ways to make a positive impact is to spread the oyster culture knowledge in this book.

Teach others about shellfish and the reasons why we should all be supporting the shellfish industry. Teach people about how shellfish are a better protein choice, or how shellfish clean the environment and create improved ecosystems, or how shellfish combat climate change by sequestering carbon, or how for every one dollar generated on a shellfish farm, three dollars are created in the local economy. Take a tour of the local shellfish farms in your area. Get to know the people who grow your food. The more people who support the shellfish industry, the larger it can grow, and in turn, the more people and animals it will support. The industry needs your help.

My friends at Island Creek Oysters have taken the "sharing is caring" approach to a new level; they are building hatcheries in developing countries and training the locals to farm their own shellfish.

Through a partnership between Dr. Hauke Kite-Powell at the Woods Hole Oceanographic Institution and the Island Creek Oysters Foundation (ICOF), a shellfish hatchery was launched in Zanzibar in the rural village of Bububu, where locals were trained on the methods of clam propagation and cultivation with the aim of not only supplying the village with sustenance foods, but also creating surplus that could be sold.

In Haiti, ICOF teamed up with Dr. Valentin Abe, founder of the Caribbean Harvest Foundation, to fund two fish hatcheries that support two tilapia farms, one in Lake Azuéi and the other in Lake Péligre. More than 450 families earn a living farming fish from the two lakes, which now generate considerable income for the local communities.

In 2018, a group of shellfish growers partnered with the Nature Conservancy to create the Shellfish Growers Climate Coalition. With more than 250 businesses from twenty-five states and Canada on board, the goal is to advocate for climate policy and showcase how the industry is making strides for a more sustainable future. As a board member of the organization, I can attest to the change we've enacted and projects we've witnessed. From shell recycling programs and oyster reef restoration to hatcheries developing ocean-acidification-resistant oyster strains, companies making compostable mesh bags for the industry, and individuals creating blogs about merroir and the benefits of eating shellfish—oyster culture is alive and spreading.

The FAO lays it out bluntly:

The high and growing prevalence of hunger and malnutrition in all its forms in the world, combined with climate and

*environmental concerns, suggests that the global food system
is failing to deliver safe, nutritious, sustainable and equitable
diets. As a result, the international community is calling for
a transformation of food systems as highlighted at the 2021
UN Food Systems Summit.*

*At the same time, population growth and rising afflu-
ence are fueling demand for more food and for resource
intensive diets. In this landscape of demand and need,
visions of what constitutes progress towards a sustainable
food system diverge, but most of them incorporate aquatic
foods as a vehicle to ending hunger and malnutrition and
building nature-positive, efficient, inclusive, resilient and
sustainable food systems for all. Aquatic food systems can
make key contributions to food security and nutrition, help
prepare for and buffer the impacts of climate change, and
when properly transformed sustainably increase the supply
of nutritious food and contribute to community resilience,
decent employment, equity, gender equality, and poverty
alleviation. Through Blue Transformation, aquatic food
systems can:*

> *A. support the provision of enough aquatic food for a
> growing population that is environmentally, socially
> and economically sustainable and equitable;*
> *B. ensure the availability and accessibility of safe and
> nutritious aquatic food for all, in particular for vulnera-
> ble populations, and reduce food loss and waste;*
> *C. ensure that aquatic food systems contribute to
> improving rights and income of vulnerable communities
> to achieve equitable livelihoods; and*
> *D. support resilience in aquatic food systems that are
> highly influenced by dynamic anthropogenic and*

non-anthropogenic processes, including from a chang-
ing climate.*

There is no doubt that our foods of the future will come from the ocean. However, the speed at which this transformation takes place will be the sole deciding factor in how much we can mitigate the damage from the current food production system. Make no mistake, our current system is aiding the destruction of the planet. The current food industry is unsustainable and lacking the ability to meet current and future demands. The only way sustained growth of our population is possible is if we support the foods and products that the shellfish industry produces. As demonstrated, this industry presents one of the greatest opportunities to transform our food supply and ecosystems in a positive way. The shellfish industry is one of the only food production systems on Earth that can help reverse the negative impacts of the past. With your help and support of the shellfish industry, we can continue to spread the knowledge, expand the industry, feed the future, and rebuild ecosystems and economies around the world.

---

* FAO, *Blue Transformation: Roadmap 2022–2030* (Rome: FAO, 2022), fao.org/3/cc0459en/cc0459en.pdf.

THE WORLD IS
YOUR OYSTER

HOPE MY OPTIMISM FOR THE FUTURE HAS TRANS-
ferred through these pages. A glimmering hope for the future
of humanity can be found in the pillar of this new food production source of ocean farms and aquaculture. We have the ability
today to shape one of the fastest-growing, most important food
production industries in the history of humankind. It's like watch-
ing the first seeds being planted twelve thousand years ago.

The shellfish farming industry is sure to continue its growth
trajectory into the future as new generations are just coming to
discover the magic of shellfish. The global shellfish industry was
valued at around $51 billion in 2021 and is projected to reach $66
billion by 2030, growing at an average compound annual growth
rate of 3 percent. As great as this sounds, it's not fast enough.

Globally, I believe we will see the shellfish industry develop
quickest in developing nations, as demand for sustainable farm

proteins and sustenance rises with the growing population. Piggybacking off the infrastructure and hatchery breakthroughs in aquaculture-developed countries, these developing nations have the ability to scale quickly and efficiently, lifting local economies out of poverty and providing food security. Yet this won't be enough. We need developed countries, such as the U.S. and those in the EU, to replace their protein choices of today with an increase in shellfish consumption.

The rise in global ocean levels will create one positive effect: as low-lying areas become inundated with ocean waters, we can and should quickly transform them into shellfish farms. Yes, land will be lost, but we should instead look at this loss as a gain in farmable oceanfront. Rising oceans will expand shellfish aquaculture real estate across the world. The glass is always half full.

Oysters are the hardiest of all marine species, but expect to see many more varieties of shellfish finding their places on the farms of the future. Seaweed farming is the newest crop to appear on farms along both coasts of the United States, providing the same ecological and nutritional benefits as shellfish, with the same minimal resource footprint. In the near future, as our ocean farms begin to outproduce the demand for human consumption, we will begin to see these products fill as many niches as possible within other industries, especially as a fishmeal and animal feed replacement.

The understanding of merroir is another aspect of the industry that is likely to grow in aquaculture-developed countries. Like wine sommeliers, there are now "oyster sommeliers," folks who travel the world tasting oysters and describing the nuanced flavor differences. The environmental attributes that contribute to merroir have yet to be fully explored, and when fully discovered will undoubtedly lead to breakthroughs in flavor profiles and production standards. The understanding of merroir may be able to teach us how to grow more nutritional marine species and aid in creating breeding programs that focus on faster growing attributes, faster filtering attributes, and resistance to ocean acidification.

As farms scale up to meet demand, we will inevitably see the consolidation of farms under umbrella companies (conglomerates of farms, all individually owned, yet working together). These "corporate" shellfish companies may become publicly listed for trading, allowing mammoth amounts of investment into the industry that is currently unavailable.

Every productive coastline around the world will eventually contain a myriad of aquaculture farms, providing nearby cities with fresh farmed seafood products year-round. Boat yards, infrastructure, and other marine services that support the farm fleets and workers will begin to thrive, creating additional jobs and opportunities. The entire blue economy will be aligned to support the aquaculture sector as our food production systems transition to aquaculture products as the baseline.

We are just a few small breakthroughs away from grasping control of our oceans and converting them into the most productive regions on Earth. Slowly but surely, we are returning to the oceans that originally sustained us and allowed our species to thrive.

The future of food is blue.

ILLUSTRATION AND PHOTO CREDITS

PAGE   8   Dan Martino, *Shellfish Species; Mussel, Scallop, Oyster, Clam.*

PAGE   14   P. S. Galtsoff, "The American Oyster, *Crassostrea virginica* Gmelin," *Fishery Bulletin* 64 (1964): 67.

PAGE   61   John Singer Sargent, *The Oyster Gathers [sic] of Cancale*, Detroit Publishing Co., between 1900 and 1920, archived in the Library of Congress Web Archives at loc.gov/item/2016817234/.

PAGE   81   Albert Berghaus, *Oyster Stalls and Lunch Rooms at Fulton Market, Fulton Street, N.Y.*, in Frank Leslie's Illustrated Newspaper, May 18, 1867, archived in the Library of Congress Web Archives at hdl.loc.gov/loc.pnp/cph.3c28743.

PAGE   82   Alfred R. Waud, *Oyster Stands in Fulton Market, in Harper's Weekly*, October 29, 1870, archived in the Library of Congress Web Archives at loc.gov/item/2001695512.

PAGE   84   Reginald Hotchkiss, *Oyster Tongers, Rock Point, Maryland*, 1941, archived in the Library of Congress Web Archives at loc.gov/item/2017745202.

PAGE   87   *Oyster Pile, Hampton, Va.*, Detroit Publishing Co., between 1900 and 1920, archived in the Library of Congress Web Archives at loc.gov/item/2016795954/.

PAGE   91   Schell and Hogan, *The Oyster War in Chesapeake Bay, in Harper's Weekly*, March 1, 1884, archived in the Library of Congress Web Archives at loc.gov/pictures/item/2002698358/.

PAGE   96   Lewis Hine, *All these work in Peerless Oyster Co. Had to get the photo while bosses were at dinner, as they refused to permit children in the photos*, March 1911, Bay St. Louis, Mississippi, archived in the

Library of Congress Web Archives at loc.gov/item/2018676343.

PAGE   96   Lewis Hine, *Rosy, an eight-year-old oyster shucker who works steady all day from about 3:00 A.M. to about 5 P.M. in Dunbar Cannery,* March 1911, Dunbar, Louisiana, archived in the Library of Congress Web Archives at loc.gov/item/2018676358.

PAGE   97   *Unloading Oyster Luggers, Baltimore, Md.,* 1905, Detroit Publishing Co., archived in the Library of Congress Web Archives at loc.gov/item/2016794933.

PAGE   98   Currier and Ives, *Honest Abe Taking Them on the Half Shell,* 1860, archived in the Library of Congress Web Archives at loc.gov/item/2003674568.

PAGE   127   Baby oysters being planted on the Cottage City oyster farm.

PAGE   156   Cottage City oyster with homemade mignonette sauce.

PAGE   157   Shucking a Cottage City oyster at the raw bar.

PAGE   158   The open-ocean Cottage City oyster farm off the coast of Oak Bluffs, Martha's Vineyard, Massachusetts.

PAGE   158   Cottage City Oysters bottom cage hoisted to the surface for harvest and an air-dry.

PAGE   226   Hannah Ritchie, "50% of All Land in the World Is Used to Produce Food," World Economic Forum, December 11, 2019, weforum.org/agenda/2019/12/agriculture-habitable-land.

PAGE   269   United Nations, "Food and Climate Change: Healthy Diets for a Healthier Planet," un.org/en/climatechange/science/climate-issues/food.

REFERENCES

Adams, Connie. "Taylor Shellfish Farms." Seattle Dining. February 2017. seattledining.com/Archive/food_products/taylor_shellfish_part_1_2017.aspx.

"Algae, Phytoplankton and Chlorophyll." Fondriest Environmental. October 22, 2014. fondriest.com/environmental-measurements/parameters/water-quality /algae-phytoplankton-chlorophyll/.

Annual Report of the Commissioners of Shell Fisheries. Providence: The Oxford Press, 1921.

Antaya, Nicole and André F. Brito. "Results from Short-Term Studies Using Kelp Meal as a Supplement to Dairy Animals at the University of New Hampshire." NODPA. October 16, 2018. nodpa.com/n/244/Results-from-Short-Term-Studies-Using-Kelp-Meal-as-a-Supplement-to-Dairy-Animals-at-the-University-of-New-Hampshire.

"Archaeological Shell Midden at Weipa." National Archives of Australia. naa.gov.au/learn/learning-resources/learning-resource-themes/first-australians /history/archaeological-shell-midden-weipa.

Arsenault, Chris. "Only 60 Years of Farming Left If Soil Degradation Continues." *Scientific American.* December 5, 2014. scientificamerican.com/article/ only-60-years-of-farming-left-if-soil-degradation-continues.

Balukjian, Brad. "Ancient Shellfish Suggest Modern Humans Evolved 50,000 Years Ago." *LA Times.* June 18, 2013. latimes.com/science/sciencenow/la-sci-sn-ancient-shellfish-human-evolution-20130618-story.html.

Benton, Michael. "Wipeout: When Life Nearly Died." New Scientist. April 26, 2003. newscientist.com/article/mg17823925-000-wipeout-when-life-nearly-died/.

Boissoneault, Lorraine. "Skulls With 'Surfer's Ear' Suggest Ancient Pearl Divers in Panama." Smithsonian Magazine. January 3, 2019. smithsonianmag. com/science-nature/bony-growths-ear-canal-plagued-pre-columbian-divers-panama-180971164/.

Bosly-Pask, Nick. "The Angasi Oyster Makes a Comeback." ABC Rural. April 10, 2017. abc.net.au/news/rural/2017-04-10/the-return-of-the-angasi-oyster/8426178.

Botta, Robert, Frank Asche, J. Scott Borsum, and Edward V. Camp. "A Review of Global Oyster Aquaculture Production and Consumption." *Marine Policy* 117 (July 2020). doi.org/10.1016/j.marpol.2020.103952.

Bricker, Suzanne B., Raymond E. Grizzle, Philip Trowbridge, Julie M. Rose, Joao G. Ferreira, Katharine Wellman, Changbo Zhu et al. "Bioextractive Removal of Nitrogen by Oysters in Great Bay Piscataqua River Estuary, New Hampshire, USA." *Estuaries and Coasts* 43 (2020): 23–38. doi.org/10.1007/s12237-019-00661-8.

Brooks, William K. *The Oyster: A Popular Summary of a Scientific Study*. Baltimore: Johns Hopkins, 1891.

Bruns, H. Arnold. "Southern Corn Leaf Blight: A Story Worth Retelling." *Agronomy* 109, no. 4 (2017): 1–7. doi.org/10.2134/agronj2017.01.0006.

Buestel, Dominique, Michel Ropert, Jean Prou, and Goulletquer Philippe. "History, Status, and Future of Oyster Culture in France" *Journal of Shellfish Research* 28, no. 4 (December 2009). doi.org/10.2983/035.028.0410.

Cecelski, David. "A World Built of Oyster Shells." February 7, 2019. davidcecelski.com/2019/02/07/a-world-built-of-oyster-shells/.

"Celebrate World Oyster Day with Australia's Oyster Coast: Exploring Three Unique Oyster Species." Australia's Oyster Coast. August 1, 2023. australiasoystercoast.com/blog/celebrate-world-oyster-day.

Chun, Beth, Gonzalo Mon, and Paul Singer. "NY Attorney General Sues JBS Over Greenwashing." JD Supra. March 11, 2024. jdsupra.com/legalnews/ny-attorney-general-sues-jbs-over-5209492/.

Chung, Laura. "Why Your Christmas Lunch Is Under Threat from an Oyster Killer." *Sydney Morning Herald*. September 5, 2022. smh.com.au/environment/sustainability/disease-kills-entire-port-stephens-sydney-rock-oyster-crop-and-farmers-incomes-20220904-p5bf7o.html.

Clark, Michael A., Nina G. G. Domingo, Kimberly Colgan, Sumil K. Thakrar, David Tilman, John Lynch, Inês L. Azevedo, and Jason D. Hill. "Global Food System Emissions Could Preclude Achieving the 1.5° and 2°C Climate Change Targets." *Science* 370, no. 6517 (November 6, 2020): 705–708. doi.org/10.1126/science.aba7357.

Cohan, Michelle. "Nemo's Garden: The Future of Farming Could Be Under the Sea." CNN. April 27, 2023. cnn.com/travel/article/nemos-garden-underwater-farm-italy-spc-intl/index.html.

Coleman, K. M. "The Lucrine Lake at Juvenal 4.141." *Classical Quarterly* 44, no. 2 (1994): 554–57. jstor.org/stable/639665.

Committee on a Framework for Assessing the Health, Environmental, and Social Effects of the Food System. "Annex 1, Dietary Recommendations for Fish Consumption." In *A Framework for Assessing Effects of the Food System*, edited by Malden C. Nesheim, Maria Oria, and Peggy Tsai Yih. Washington, DC: National Academies Press, 2015. ncbi.nlm.nih.gov/books/NBK305180/.

Conte, Fred S. "California Oyster Culture." *California Aquaculture* [UC Davis]. marine-aquaculture.extension.org/wp-content/uploads/2019/05/California-Oyster-Culture.pdf.

"COP26: Agricultural Expansion Drives Almost 90 Percent of Global Deforestation." FAO. June 11, 2021. fao.org/newsroom/detail/cop26-agricultural-expansion-drives-almost-90-percent-of-global-deforestation/en.

"Covid Brings Lasting Change to the French Oyster Market." Eurofish International Organisation. February 23, 2023. eurofish.dk/covid-brings-lasting-change-to-the-french-oyster-market/.

Cozzi, Laura, Olivia Chen, and Hyeji Kim. "The World's Top 1% of Emitters Produce Over 1000 Times More CO2 Than the Bottom 1%." International Energy Agency. iea.org/commentaries/the-world-s-top-1-of-emitters-produce-over-1000-times-more-co2-than-the-bottom-1.

Crook, Steven. "Oyster Shells and Breakwaters: Taiwan's Fishing Villages." Taiwan Business TOPICS. July 24, 2020. topics.amcham.com.tw/2020/07/taiwans-fishing-villages/.

Davenport, Coral. "Strawberry Case Study: What If Farmers Had to Pay for Water?" *New York Times*. December 29, 2023. nytimes.com /interactive/2023/12/29/climate/california-farmers-water-tax.html.

Deaton, Jeremy. "Could Underwater Farming Feed the World?" Popular Science. December 5, 2016. popsci.com/3d-underwater-farming/.

"Department of Natural Resources." Maryland Manual On-Line. msa.maryland.gov/msa/mdmanual/21dnr/html/dnrf.html.

DeSmog. "World's Largest Meat Company, JBS, Increases Emissions in Five Years Despite 2040 Net Zero Climate Target, Continues to Greenwash Its Huge Climate Footprint." Institute for Agriculture and Trade Policy. April 21, 2022. iatp.org/media-brief-jbs-increases-emissions-51-percent.

Dorey, Fran. "When Did Modern Humans Get to Australia?" Australian Museum. September 12, 2021. australian.museum/learn/science/human-evolution/the-spread-of-people-to-australia/.

"Eastern Oysters." Chesapeake Bay Foundation. cbf.org/about-the-bay /chesapeake-wildlife/eastern-oysters/index.html.

Edmond, Charlotte. "New York Is Building a Wall of Oysters to Fight Flooding." World Economic Forum. December 14, 2021. weforum.org/agenda/2021/12/ new-york-oysters-sea-rise-climate-change/.

"1872 Oyster Farming Begins in Sydney." Australian Food Timeline. australianfoodtimeline.com.au/oyster-farming-begins/.

"Energy Use in Food Production." Choose Energy. November 26, 2019. chooseenergy.com/blog/energy-101/energy-food-production.

EUMOFA. *Oysters in the EU: Price Structure in the Supply Chain*. Luxembourg: EU, 2022. eumofa.eu/documents/20178/517783/PTAT+Oyster_EN .pdf/385bcceb-50f5-4f80-04a1-227416522434?t=1668076669628.

Eyton, Thomas Campbell. *A History of the Oyster and the Oyster Fisheries*. London: John Van Voorst, 1858. google.com/books/edition/A_History_of_the_ Oyster_and_the_Oyster_F/jU1DAAAAIAAJ?hl=en&gbpv=1.

FAO. *Agriculture, Forestry and Other Land Use Emissions by Sources and Removals by Sinks*. FAO, 2014. fao.org/3/i3671e/i3671e.pdf.

FAO. *Blue Transformation: Roadmap 2022–2030*. Rome: FAO, 2022. fao.org/3/cc0459en/cc0459en.pdf.

FAO. *How to Feed the World in 2050*. FAO, 2009. fao.org/fileadmin/templates/wsfs/docs/expert_paper/How_to_Feed_the_World_in_2050.pdf.

FAO. *The State of World Fisheries and Aquaculture 2018: Meeting the Sustainable Development Goals*. Rome: FAO, 2018. fao.org/3/i9540en/I9540EN.pdf.

FAO. *The State of World Fisheries and Aquaculture 2022: Towards Blue Transformation*. Rome: FAO, 2022. doi.org/10.4060/cc0461en.

"Farmers Harvest Oysters at Oyster Cultivation Fields in Qinzhou, S China's Guangxi." Xinhua. July 22, 2019. xinhuanet.com/english/2019-07/22/c_138246009_6.htm.

Fava, Marta. "The Threats to the Ocean in 2022 and How to Prevent Them." UNESCO Intergovernmental Oceanographic Commission. May 9, 2022. oceanliteracy.unesco.org/threats-to-the-ocean/.

Feng, Jing-Chun, Liwei Sun, and Jinyue Yan. "Carbon Sequestration via Shellfish Farming: A Potential Negative Emissions Technology." *Renewable and Sustainable Energy Reviews* 171 (January 2023). doi.org/10.1016/j.rser.2022.113018.

Finchman, Michael W. "Trials & Errors & Triploids: Odyssey of an Oyster Inventor." *Chesapeake Quarterly* 9, no. 2 (June 2010). chesapeakequarterly.net/V09N2/main2/.

"Food Choices and the Planet." EarthSave International. earthsave.org/environment.htm.

"For a Clean Chesapeake Bay." Chesapeake Oyster Alliance. chesapeakeoysteralliance.org/index.html.

Fox, Richard. "*Crassostrea virginica*." Invertebrate Anatomy OnLine. May 25, 2007. lanwebs.lander.edu/faculty/rsfox/invertebrates/crassostrea.html.

Fujiya, M. "Oyster Farming in Japan." *Helgolander Wiss. Meeresunters* 20 (1970): 464–479. hmr.biomedcentral.com/track/pdf/10.1007/BF01609922.pdf.

Galtsoff, P. S. "The American Oyster, *Crassostrea virginica* Gmelin," *Fishery Bulletin* 64 (1964): 67.

Gapps, Stephen. "An Australian Stonehenge?" Sea Museum. January 11, 2017. sea.museum/2017/01/11/an-australian-stonehenge.

Gentry, Rebecca R., Halley E. Froehlich, Dietmar Grimm, Peter Kareiva, Michael Parke, Michael Rust, Steven D. Gaines, and Benjamin S. Halpern. "Mapping the Global Potential for Marine Aquaculture." *Nature Ecology & Evolution* 1 (2017): 1317–1324. doi.org/10.1038/s41559-017-0257-9.

Gephart, Jessica A., Patrik J. G. Henriksson, Robert W. R. Parker, Alon Shepon, Kelvin D. Gorospe, Kristina Bergman, Gidon Eshel et al. "Environmental Performance of Blue Foods." *Nature* 597 (September 15, 2021): 360–365. doi.org/10.1038/s41586-021-03889-2.

Gillis, Anna Maria. "Oyster Wars: Wayward Watermen of the Chesapeake Bay." *Humanities* 32, no. 3 (May/June 2011). neh.gov/humanities/2011/mayjune/statement/oyster-wars.

Godfrey, Mark. "Home Province of Chinese Processing Port Dalian Sees Jump in Imports, Value." Seafood Source. seafoodsource.com/news/supply-trade/home-province-of-chinese-processing-dalian-sees-jump-in-imports-value.

Gordon, Arielle. "The Centuries-Long Saga of the 'Oyster Wars.'" Boundary Stones. November 18, 2020. boundarystones.weta.org/2020/11/18/centuries-long-saga-%E2%80%98oyster-wars%E2%80%99.

Gould, Augustus. *Invertebrata of Massachusetts*. Cambridge, MA: Folsom, Wells, and Thurston, 1841.

Grizel, Henri, and Maurice Héral. "Introduction into France of the Japanese Oyster (*Crassostrea gigas*)." *J. Cons. Int. Explor. Mer* 47 (1991): 399–403. archimer.ifremer.fr/doc/1991/publication-2760.PDF.

Groesbeck, Amy S., Kristen Rowell, Dana Lepofsky, and Anne K. Salomon. "Ancient Clam Gardens Increased Shellfish Production: Adaptive Strategies from the Past Can Inform Food Security Today." *PLOS One* 9, no. 3 (March 2014). doi.org/10.1371/journal.pone.0091235.

"Hangtown Fry History and Recipe." What's Cooking America. whatscookingamerica.net/history/hangtownfryhistory.htm.

Hautmann, Michael, David Ware, and Hugo Bucher. "Geologically Oldest Oysters Were Epizoans on Early Triassic Ammonoids." *Journal of Molluscan Studies* 83, no. 3 (August 2017): 253–260. doi.org/10.1093/mollus/eyx018.

Hayes, Clara. "What Nutrients Are in Marine Phytoplankton?" Plankton for Health. planktonforhealth.co.uk/what-nutrients-are-in-marine-phytoplankton/.

Hill, Holly. "Food Miles: Background and Marketing." National Center for Appropriate Technology. 2008. attra.ncat.org/publication/food-miles-background-and-marketing/.

"History of Oyster Farming in Wallis Lake." Barclay Oysters. barclayoysters.com.au/history.

Homer. *The Iliad.* Translated by A. T. Murray. Cambridge, MA: Harvard University Press; London: William Heinemann; 1924.

Hunter, Mitch, Ann Sorensen, Theresa Nogeire-McRae, Scott Beck, Stacy Shutts, and Ryan Murphy. *Farms Under Threat 2040: Choosing an Abundant Future.* Washington, DC: American Farmland Trust, 2022. farmlandinfo.org/publications/farms-under-threat-2040/.

Imbler, Sabrina. "Found: 588 Carp Teeth From China's Oldest-Known Fish Farm." Atlas Obscura. October 1, 2019. atlasobscura.com/articles/carp-teeth-neolithic-chinese-aquaculture.

"Indoor Water Use at Home." Water Footprint Calculator. Updated July 15, 2022. watercalculator.org/footprint/indoor-water-use-at-home/.

Ingersoll, Ernest. *The Oyster Industry.* Washington: Government Print Office, 1881.

"Interest in Shellfish Aquaculture Leads to Misconceptions About Triploid Oysters." North Carolina Department of Environmental Quality. May 2018.

deq.nc.gov/about/divisions/marine-fisheries/news-media/insight-newsletter/may-2018/interest-shellfish-aquaculture-leads-misconceptions-about-triploid-oysters.

IPCC. *Climate Change 2014: Mitigation of Climate Change: Working Group III Contribution to the Fifth Assessment Report of the Intergovernmental Panel on Climate Change*. Edited by Ottmar Edenhofer, Ramon Pichs-Madruga, Youba Sokona, Jan C. Minx, Ellie Farahani, Susanne, Kadner, Kristin Seyboth, et al. New York: Cambridge University Press, 2014.

Isa, Mari. "The Great Oyster Craze: Why 19th Century Americans Loved Oysters." MSU Campus Archaeology Program. February 23, 2017. campusarch.msu.edu/?p=4962.

Jain, Atul, and Xiaoming Xu. "Food Production Generates More Than a Third of Manmade Greenhouse Gas Emissions – A New Framework Tells Us How Much Comes from Crops, Countries and Regions." The Conversation. September 13, 2021. theconversation.com/food-production-generates-more-than-a-third-of-manmade-greenhouse-gas-emissions-a-new-framework-tells-us-how-much-comes-from-crops-countries-and-regions-167623.

Jian, Lee. "Forget Fine Dining, Korean Winters Offer Oysters a Plenty." *Korea JoongAng Daily*. January 9, 2023. koreajoongangdaily.joins.com/2023/01/09/culture/foodTravel/Korea-Seoul-Oysters/20230109162510551.html.

Jones, Prudence. "Cleopatra's Cocktail." *Classical World* 103, no. 2 (Winter 2010): 207–220. academia.edu/1137049/Cleopatras_Cocktail_Classical_World_103_2_2010_207_220.

Joseph, Michael. "9 Health Benefits of Oysters (and Full Nutrition Facts)." Nutrition Advance. September 26, 2023. nutritionadvance.com/oysters-nutrition-benefits/.

Kadner, Kristin Seyboth, et al. New York: Cambridge University Press, 2014. ipcc.ch/site/assets/uploads/2018/02/ipcc_wg3_ar5_full.pdf.

Kazarian, Christopher. "New England Shellfish Growers Plant Seeds for Better Lives at Home and Abroad." Global Seafood Alliance. February 22, 2016. globalseafood.org/advocate/social-oysters-aquaculture-inspiring-communities/.

Kennedy, Victor S., and Linda L. Breisch. *Maryland's Oysters: Research and Management*. Maryland Sea Grant, 2001. chesapeakequarterly.net/images/uploads/siteimages/imported/mdoyst3.pdf.

Kinsley, Natalie. "Mussel Waste Fed to Hens for Nutrition and the Environment." Poultry World. July 12, 2021. poultryworld.net/poultry/mussel-waste-fed-to-hens-for-nutrition-and-the-environment/.

"Latest Western Cape Dam Levels." Western Cape Government. westerncape.gov.za/general-publication/latest-western-cape-dam-levels.

Le Gal, Yves. "2009: The Concarneau Marine Biology Laboratory Celebrates Its 150th Anniversary." *La Lettre du Collège de France* 4 (2009). doi.org/10.4000/lettre-cdf.779.

Lewis, Dan. "The Pearly History Behind Chinese Takeout Boxes." Now I Know. February 8, 2022. nowiknow.com/the-pearly-history-behind-chinese-takeout-boxes/.

Lockhart, Emma. "Phoenix Approves Water Conservation Ordinance, Impacting Businesses That Use 250,000 Gallons or More Per Day." Arizona Family. March 7, 2024. azfamily.com/2024/03/08/phoenix-approves-water-conservation-ordinance-impacting-businesses-that-use-250000-gallons-or-more-per-day/.

Lockwood, Samuel. "The Natural History of the Oyster I." *Popular Science Monthly* 6 (November 1874). en.wikisource.org/wiki/Popular_Science_Monthly/Volume_6/November_1874/The_Natural_History_of_the_Oyster_I.

Lockwood, Samuel. "The Natural History of the Oyster II." *Popular Science Monthly* 6 (December 1874). en.wikisource.org/wiki/Popular_Science_Monthly/Volume_6/December_1874/The_Natural_History_of_the_Oyster_II.

MacKenzie, Clyde L. "History of Oystering in the United States and Canada, Featuring the Eight Greatest Oyster Estuaries" *Marine Fisheries Review* 58, no. 4 (1996). spo.nmfs.noaa.gov/sites/default/files/pdf-content/MFR/mfr584/mfr5841.pdf.

Mao, Yuze, Fan Lin, Jianguang Fang, Jinghui Fang, Jiaqi Li, and Meirong Du. (2019). "Bivalve Production in China." In *Goods and Services of Marine Bivalves*.

Edited by Aad C. Smaal, Joao G. Ferreira, Jon Grant, Jens K. Petersen, and Øivind Strand. Springer, 2018. doi.org/10.1007/978-3-319-96776-9_4.

March, Janine. "Oysters, the French Obsession." The Good Life France. thegoodlifefrance.com/oysters-the-french-obsession/.

Marean, Curtis W. "When the Sea Saved Humanity." Scientific American. November 1, 2012. scientificamerican.com/article/when-the-sea-saved-humanity-2012-12-07/.

"Marennes-Oléron Oysters PGI." Taste France. tastefrance.com/us/french-products/seafood/marennes-oleron-oysters-pgi.

Marshall, Michael. "Timeline: The Evolution of Life." New Scientist. Updated April 27, 2023. newscientist.com/article/dn17453-timeline-the-evolution-of-life/.

Marzano, Annalisa. *Harvesting the Sea: The Exploitation of Marine Resources in the Roman Mediterranean*. Oxford, U.K.: Oxford University Press, 2013.

Maximus, Valerius. *Memorable Deeds and Sayings: One Thousand Tales from Ancient Rome*. Indianapolis: Hackett Publishing, 2004.

Minkel, J. R. "Earliest Known Seafood Dinner Discovered." *Scientific American*. October 17, 2007. scientificamerican.com/article/earliest-known-seafood-di/.

Mon, Gonzalo E., and Katie Rogers. "NARB Agrees Advertiser Can't Support Aspirational Net Zero Claims." Kelley Drye. June 22, 2023. kelleydrye.com/viewpoints/blogs/ad-law-access/narb-agrees-that-advertiser-cant-support-aspirational-net-zero-claims.

Montaigne, Fen. "The Fertile Shore." *Smithsonian Magazine*. January 2020. smithsonianmag.com/science-nature/how-humans-came-to-americas 180973739/.

Moss, Madonna L., and Jon M. Erlandson. "Diversity in North Pacific Shellfish Assemblages: The Barnacles of Kit'n'Kaboodle Cave, Alaska." *Journal of Archaeological Science* 37, no. 12 (December 2010): 3359–3369. doi.org/10.1016/j.jas.2010.08.004.

Murray, Eustace Clare Grenville. *The Oyster: Where, How, and When to Find,*

Breed, Cook, and Eat It. London: Trubner & Co., 1861.

National Academy of Sciences, National Academy of Engineering, and Institute of Medicine. *Population Summit of the World's Scientific Academies*. Washington, DC: The National Academies Press, 1993. doi.org/10.17226/9148.

Nigro, Carmen. "History on the Half-Shell: The Story of New York City and Its Oysters." New York Public Library. June 2, 2011. nypl.org/blog/2011/06/01/ history-half-shell-intertwined-story-new-york-city-and-its-oysters.

"Nihojima-mura (Nijo Island Village)." nehori.com/niho/e.

Nobbs, Jeff. "Why Everyone Should Eat Oysters." Jeff Nobbs [blog]. June 24, 2021. jeffnobbs.com/posts/why-everyone-should-eat-oysters.

Oberg, Kalervo. "South American Nomad." Encyclopædia Britannica. January 25, 2007. britannica.com/topic/South-American-nomad.

Ólafsson, Björn. "What Monoculture Farming Is, and Why It Matters." Sentient Media. September 27, 2023. sentientmedia.org/monoculture.

"On the Forefront of Nutrient Credit Trading Using Oysters: Lessons Learned." NOAA. coast.noaa.gov/digitalcoast/training/oyster-aquaculture.html.

"193 Countries in the United Nations Approve Treaty to Stop the Oceans from Dying." Beyond Pesticides. March 10, 2023. beyondpesticides.org/ dailynewsblog/2023/03/193-countries-in-the-united-nations-approve-treaty- to-stop-the-oceans-from-dying/.

"Our Industry." Oysters Tasmania. oysterstasmania.org/ourindustry.html.

"Oyster Culture in Taiwan." Taiwan Panorama. April 1979. taiwanpanorama .com/en/Articles/Details?Guid=08e8689d-58c5-4771-93bb-3e07e6fcac56&- CatId=9&postname=Oyster%20Culture%20in%20Taiwan.

"Oyster Harvest in Dalian, NE China's Liaoning." Xinhua. September 24, 2019. xinhuanet.com/english/2019-09/24/c_138417822_10.htm.

"Oyster Industry in NSW." NSW Department of Primary Industries. dpi.nsw .gov.au/fishing/aquaculture/publications/oysters/oyster-industry-in-nsw.

"Oysters in Japan – Everything You Need to Know." Gurunavi. February 22, 2017.   gurunavi.com/en/japanfoodie/2017/02/oysters-in-japan.html?__ngt__=TT145b4c400006ac1e4ae4b5Z9bQ_GXbH0Q6UV2XXjTQSI.

"Oysters on the Rise." Chesapeake Bay Foundation. January 26, 2023. cbf.org/blogs/save-the-bay/2023/01/oysters-on-the-rise.html.

"Oysters—Sources, Quantities and Cultivation Methods." Seafish. November 20, 2020. seafish.org/responsible-sourcing/aquaculture-farming-seafood/species-farmed-in-aquaculture/aquaculture-profiles/oysters/sources-quantities-and-cultivation-methods/.

"Pacific Ocean Acidification Project." NOAA. ioos.noaa.gov/project/pacific-ocean-acidification-project.

Parsons, P. J. "The Oyster." *Zeitschrift für Papyrologie und Epigraphik* 24 (1977): 1–12. jstor.org/stable/20181273.

Peralta, Ernestina M., Barry Leonard M. Tumbokon, and Augusto E. Serrano, Jr. "Use of Oyster Processing Byproduct to Replace Fish Meal and Minerals in the Diet of Nile Tilapia *Oreochromis niloticus* Fry." *Israeli Journal of Aquaculture – Bamidgeh* 68 (2016). evols.library.manoa.hawaii.edu/server/api/core/bitstreams/b61f044c-8b37-4dd8-8abe-bd1f8ac1c181/content.

Polo, Marco. *The Travels of Marco Polo*, ed. Manuel Komroff, New York: Boni & Liveright, 1926.

Potts, Billy. "What Will Save Hong Kong's 700-Year-Old Oyster Farms?" Zolima City Magazine. September 21, 2021. zolimacitymag.com/what-will-save-hong-kong-700-year-old-oyster-farms-deep-bay/.

"Q. Is There Still Radiation in Hiroshima and Nagasaki?" City of Hiroshima. city.hiroshima.lg.jp/site/english/9809.html.

Ray, Nicholas E., Timothy J. Maguire, Alia N. Al-Haj, Maria C. Henning, and Robinson W. Fulweiler. "Low Greenhouse Gas Emissions from Oyster Aquaculture." *Environmental Science & Technology* 53, no. 15 (July 11, 2019): 9118–9127. doi.org/10.1021/acs.est.9b02965.

Rice, Michael A. "A Brief History of Oyster Aquaculture in Rhode Island," in

Aquaculture in Rhode Island: 2006 Yearly Status Report. Coastal Resources Management Council, 2006. crmc.ri.gov/aquaculture/aquareport06.pdf.

Risch, Patricia, and Christian Adlhart. "A Chitosan Nanofiber Sponge for Oyster-Inspired Filtration of Microplastics." *ACS Applied Polymer Materials* 3, no. 9 (September 2021): 4685–4694. doi.org/10.1021/acsapm.1c00799.

Ritchie, Hannah. "50% of All Land in the World Is Used to Produce Food." World Economic Forum. December 11, 2019. weforum.org/agenda/2019/12/agriculture-habitable-land.

Ritchie, Hannah. "How Is Food Insecurity Measured?" Our World in Data. April 27, 2023. ourworldindata.org/food-insecurity.

Ritchie, Hannah, and Max Roser. "Land Use." Our World in Data. September 2019. ourworldindata.org/land-use.

Ritchie, Hannah, Pablo Rosado, and Max Roser. "Environmental Impacts of Food Production." Our World in Data. 2022. ourworldindata.org/environmental-impacts-of-food.

Ritchie, Hannah, Pablo Rosado, and Max Roser. "Hunger and Undernourishment." Our World in Data. 2023. ourworldindata.org/hunger-and-undernourishment.

Rojanasakul, Mira, Christopher Flavelle, Blacki Migliozzi, and Eli Murray. "America Is Using Up Its Groundwater Like There's No Tomorrow." *New York Times*. August 28, 2023. nytimes.com/interactive/2023/08/28/climate/groundwater-drying-climate-change.html.

"Seaweed Aquaculture for Food Security, Income Generation and Environmental Health in Tropical Developing Countries." World Bank Group. July 1, 2016. doi.org/10.1596/24919.

Shakespeare, William. *The Merry Wives of Windsor*. Washington, DC: Folger Shakespeare Library, n.d.

Sharma, Shefali. "The Great Climate Greenwash: Global Meat Giant JBS' Emissions Leap by 51% in Five Years." Institute for Agriculture and Trade Policy. April 20, 2022. iatp.org/jbs-emissions-rising-despite-net-zero-pledge.

"6 of the Most Sustainable Foods in the World." Nutritics. nutritics.com/en/resources/blog/6-of-the-most-sustainable-foods-in-the-world.

Smith, Kiona N. "Aquaculture May Be the Future of Seafood, But Its Past Is Ancient." Ars Technica. September 17, 2019. arstechnica.com/science/2019/09/fish-farming-may-be-much-older-than-we-thought/.

"South Korea's Booming Oyster Farming Industry Harnesses Vast Tidal Flats." La Prensa Latina. January 20, 2023. laprensalatina.com/south-koreas-booming-oyster-farming-industry-harnesses-vast-tidal-flats/.

Springmann, Marco, Daniel Mason-D'Croz, Sherman Robinson, Keith Wiebe, H. Charles J. Godfray, Mike Rayner, and Peter Scarborough. "Health-Motivated Taxes on Red and Processed Meat: A Modelling Study on Optimal Tax Levels and Associated Health Impacts." *PLOS One* 13, no. 11 (November 6, 2018). doi.org/10.1371/journal.pone.0204139.

"State Oyster Police Force." Maryland Department of Natural Resources. March 30, 1868. dnr.maryland.gov/documents/OysterPoliceAct_1868.pdf.

Steele, E. N. *The Immigrant Oyster (Ostrea gigas) Now Known as the Pacific Oyster.* Olympia, WA, 1964. wsg.washington.edu/wordpress/wp-content/uploads/Immigrant-Oyster.pdf.

Supplement, Containing the Acts Passed at a General Assembly of the Commonwealth of Virginia, of a Public and Generally Interesting Nature, Passed Since the Session of Assembly Which Commenced in the Year One Thousand Eight Hundred and Seven. Richmond: Samuel Pleasants, 1812. babel.hathitrust.org/cgi/pt?id=osu.32437123258820&view=1up&seq=82&q1=oysters.

"The Chesapeake Bay 'Bolide' That Shaped the Groundwater in Southeastern Virginia." Virginia Places. virginiaplaces.org/geology/bolide.html.

"The Chesapeake Bay Bolide Impact: A New View of Coastal Plain Evolution." USGS. pubs.usgs.gov/fs/fs49-98/.

"The Colchester Oyster Feast." Camulos. camulos.com/oyster.htm.

"The History of the Oyster." France Naissain. francenaissain.com/en/the-oyster/the-oyster-and-its-origins/the-history-of-the-oyster/.

"The Many Reasons to Love Oysters—Even If You Hate Them." Cleveland Clinic. July 17, 2020. health.clevelandclinic.org/7-reasons-to-love-oysters-even-if-you-hate-them.

"The True Environmental Cost of Eating Meat." Connect 4 Climate. connect4climate.org/infographics/true-environmental-cost-eating-meat.

"The Ultimate Guide to the Étang de Thau." European Waterways. europeanwaterways.com/blog/etang-de-thau-guide/.

Thompson, Victor D., Torben Rick, Carey J. Garland, David Hurst Thomas, Karen Y. Smith, Sarah Bergh, Matt Sanger et al. "Ecosystem Stability and Native American Oyster Harvesting Along the Atlantic Coast of the United States." *Science Advances* 6, no. 28 (July 10, 2020). doi.org/10.1126/sciadv.aba9652.

Tollefson, Jeff. "UN Decision Puts Brakes on Ocean Fertilization." *Nature* 453, no. 7196 (June 2008). doi.org/10.1038/453704b.

Ullstrup, A. J. "The Effects of the Southern Corn Leaf Blight Epidemic of 1970–1971." *Annual Review of Phytopathology* 10 (September 1972): 37–50. doi.org/10.1146/annurev.py.10.090172.000345.

United Nations. "Food and Climate Change: Healthy Diets for a Healthier Planet." un.org/en/climatechange/science/climate-issues/food.

United Nations. "Secretary-General Warns of Climate Emergency, Calling Intergovernmental Panel's Report 'A File of Shame', While Saying Leaders 'Are Lying', Fuelling Flames." Press release no. SG/SM/21228. April 4, 2022. press.un.org/en/2022/sgsm21228.doc.htm.

United Nations. "Water Facts." unwater.org/water-facts/water-food-and-energy.

United Nations Convention to Combat Desertification. *The Global Land Outlook*. Bonn, Germany, 2017. Unccd.int/sites/default/files/documents/2017-09/GLO_Full_Report_low_res.pdf.

"U.S. Energy Facts Explained." Energy Information Administration. eia.gov/energyexplained/us-energy-facts/.

"Using Global Emission Statistics Is Distracting Us from Climate Change Solutions." CLEAR Center UC Davis. June 26, 2020. clear.ucdavis.edu/explainers/using-global-emission-statistics-distracting-us-climate-change-solutions.

Van Deelen, Grace. "Feeding Cows Seaweed Reduces Their Methane Emissions, But California Farms Are a Long Way from Scaling Up the Practice." Inside Climate News. June 14, 2022. insideclimatenews.org/news/14062022/cow-seaweed-methane/.

Vianello, Alvise, Rasmus Lund Jensen, Li Liu, and Jes Vollertsen. "Simulating Human Exposure to Indoor Airborne Microplastics Using a Breathing Thermal Manikin." *Scientific Reports* 9 (June 2019). doi.org/10.1038/s41598-019-45054-w.

Vince, Gaia. "How the World's Oceans Could Be Running Out of Fish." BBC. September 20, 2012. bbc.com/future/article/20120920-are-we-running-out-of-fish.

"Water and Agriculture." Organization for Economic Co-Operation and Development. oecd.org/agriculture/topics/water-and-agriculture/.

"Water Scarcity." World Wildlife Fund. worldwildlife.org/threats/water-scarcity.

Webber, Jemima. "$285 Billion Health Costs Could Be Curbed With Meat Taxes, Says Research." Live Kindly. livekindly.co/research-scientists-taxes-meat-combat-285-billion-health-care-costs.

Wharton, James. *The Bounty of the Chesapeake.* Charlottesville: University Press of Virginia, 1957.

"What Happens to Animal Waste?" FoodPrint. October 8, 2018, updated February 28, 2024. foodprint.org/issues/what-happens-to-animal-waste.

"What Is the Chesapeake Clean Water Blueprint?" Chesapeake Bay Foundation. cbf.org/how-we-save-the-bay/chesapeake-clean-water-blueprint/what-is-the-chesapeake-clean-water-blueprint.html.

Whitman, Levi. "A Topographical Description of Wellfleet" in *Collections of the Massachusetts Historical Society*, 3rd ed. Boston: Massachusetts Historical Society, 1794.

Wikipedia, s.v. "Calyptogena magnifica." Accessed December 19, 2023. en.wikipedia.org/w/index.php?title=Calyptogena_magnifica.

Wikipedia, s.v. "François Vatel." Accessed October 21, 2023. en.wikipedia.org/wiki/Fran%C3%A7ois_Vatel.

Wikipedia, s.v. "San Antón de Carlos." Accessed October 22, 2023. en.wikipedia.org/wiki/San_Ant%C3%B3n_de_Carlos.

Willer, David F., and David C. Aldridge. "Sustainable Bivalve Farming Can Deliver Food Security in the Tropics." *Nature Food* 1 (2020): 384–388. doi.org/10.1038/s43016-020-0116-8.

Williams, Caroline. "The Way We Were." New Scientist. July 26, 2003. newscientist.com/article/mg17924055-500-the-way-we-were/.

Yamawaka, Masaya. "The Hidden Potential of Oysters." Ministry of Agriculture, Forestry and Fisheries. artsandculture.google.com/story/the-hidden-potential-of-oysters-ministry-of-agriculture-forestry-and-fisheries/IwXR23dzCriOLQ?hl=en.

Yu, Shu, Xiaomei Hou, Changkun Huan, and Yongtong Mu. "Comments on the Oyster Aquaculture Industry in China: 1985–2020." *Thalassas* 39 (May 2023): 875–882. doi.org/10.1007/s41208-023-00558-1.

Zabarenko, Deborah. "One-Third of World Fish Catch Used for Animal Feed." Reuters. October 28, 2008. reuters.com/article/us-fish-food/one-third-of-world-fish-catch-used-for-animal-feed-idUSTRE49S0XH20081029.

Zhou, Chang-shi, Chang-kun Huan, and Yong-tong Mu. "Development Strategies of Oyster Industry in the Coastal Region of Pearl River Delta: Based on In-Depth Interviews of Mariculture Operators, Enterprises, and Government Administrative Agencies." *Research of Agricultural Modernization* 35 (2014): 757–762. doi.org/10.13872/j.1000-0275.2014.0090.

Zhu, Changbo, Paul C. Southgate, and Ting Li. "Production of Pearls." In *Goods and Services of Marine Bivalves*. Edited by Aad C. Smaal, Joao G. Ferreira, Jon Grant, Jens K. Petersen, and Øivind Strand. Springer, 2018. doi.org/10.1007/978-3-319-96776-9_5.

ACKNOWLEDGMENTS

This book wouldn't be possible without my family. Their endless support and guidance make everything possible.

ABOUT THE AUTHOR

DAN MARTINO WAS A TV CAMERAMAN/PRO-
ducer for a decade before discovering oyster farming
while filming a TV project. Learning about the culture,
environmental and economic benefits, and possible solutions to
climate change that oysters offer, Dan promptly decided to change
careers to attempt his hand at farming oysters. He sought out a
mentor and found island local Jack Blake (Sweet Neck Oysters in
Katama Bay, Edgartown, MA).

After a year mentoring on the farm with Jack, Dan was ready to
start his own farm, and with the aid of his brother Greg, he started
the Cottage City Oyster farm in Oak Bluffs in 2014.

The farm was placed in the open-ocean Vineyard Sound, unlike all of the other farms on the island, which are located in one of the ponds or bays. After two years of trials, the brothers had their first harvestable oyster. The merroir distinction was apparent immediately.

Dan's interest in the farm didn't stop with production. He soon found himself reaching out to scientists in an attempt to learn more about the ocean environment, new species to farm, and different methods of farming. Partnering with various organizations, the brothers began farming kelp seaweed in the winter of 2013 and helped craft the state regulations for the permitting of the new seaweed crop. These actions led Dan to begin serving on numerous local, state, and federal boards, including Governor Baker's Ocean Acidification Committee, the Shellfish Growers Climate Coalition, the MV Farm Bureau (serving as president), and the MA Aquaculture Association.

As the farm grew, so did Dan's education in oyster history. As a host on the Cottage City Oyster farm tour, Dan began researching the global history of oyster farming and various aspects of the modern industry. These initial findings became the impetus for this book, which was started during Covid and completed three years later.

This story is just the beginning. There are so many amazing people, styles, appellations, and journeys revolving around oysters that have yet to be told.

Thanks for reading. Spread the word.